ADVANCED
CLEANING PRODUCT FORMULATIONS

ADVANCED CLEANING PRODUCT FORMULATIONS

Household, Industrial, Automotive

by

Ernest W. Flick

Reprint Edition

NOYES PUBLICATIONS
Westwood, New Jersey, U.S.A.

Library of Congress Catalog Card Number: 89-30274
ISBN: 0-8155-1186-8

Published in the United States of America by
Noyes Publications
Fairview Avenue, Westwood, New Jersey 07675

Transferred to Digital Printing 2009

Library of Congress Cataloging-in-Publication Data

Flick, Ernest W.
     Advanced cleaning product formulations : household, industrial,
automotive / by Ernest W. Flick.
          p.   cm.
     Includes index.
     ISBN 0-8155-1186-8 :
     1. Cleaning compounds. I. Title.
TP990.F56   1989
668'.1--dc19   89-30274
                                                    CIP

Preface

This book presents more than 800 up-to-date advanced cleaning product formulations for household, industrial and automotive applications. It is the result of information received from numerous industrial companies and other organizations. The data represent selections made at no cost to, nor influence from, the makers or distributors of these materials. Only the most recent formulas have been included.

Formulation in the cleaning product industry has undergone significant change during the past few years. Raw materials costs have risen and manufacturers have been reluctant to pass along these increases. By changing formulations to improve cost/performance characteristics, manufacturers have been able to control costs but still enhance performance. This book presents manufacturers' suggested formulations which might meet these new performance criteria.

The formulations in this book are divided into the following sections and chapters:

    I.   Household/Industrial Cleaners
         1.   Bathroom Cleaners (16)
         2.   Dishwashing Detergents (57)
         3.   Disinfectants (11)
         4.   Floor Cleaners and Wax Strippers (41)
         5.   General Purpose Cleaners (73)
         6.   Laundry Products (143)
         7.   Metal Cleaners (73)
         8.   Oven Cleaners (10)
         9.   Rinse Additives and Aids (96)
       10.   Rug, Carpet and Upholstery Cleaners and Shampoos (40)
       11.   Wall and Hard Surface Cleaners (49)
       12.   Window and Glass Cleaners (25)
       13.   Miscellaneous Cleaners (129)

II.   Automotive Cleaners
    14.   Car and Truck Washes (51)
    15.   Whitewall Tire Cleaners (8)
    16.   Miscellaneous Cleaners (21)

Parenthetic numbers indicate the number of products in each chapter. Each formula is located in the chapter which is most applicable. The reader, seeking a formula for a specific end use, should check each chapter which could possibly apply. In addition to the above, there are two other sections which will be helpful to the reader:

III.   A chemical trademark section where each tradenamed raw material included in the book is listed with a chemical description and the supplier's name. The specifications which each raw material meets are included, if applicable.

IV.   Main office addresses of the suppliers of trademarked raw materials.

Each formulation in the book lists the following information, as available, in the manufacturer's own words:

- Description of end use and most outstanding properties.

- The percent by weight or volume of each raw material included in the formula, rounded to a decimal figure.

- Key properties of the formula, which are the features that the source considers to be more outstanding than other formulations of the same type.

- The formula source, which is the company or organization that supplied the formula. The secondary source may be the originating company and/or the primary source's publication title, or both. A formula number is included, if applicable.

The table of contents is organized in such a way as to serve as a subject index.

My fullest appreciation is expressed to the companies and organizations who supplied the original starting formulations included in this book. I also thank the suppliers of the raw materials included in these formulations, who furnished information describing their trademarked raw materials.

Newburyport, Massachusetts   Ernest W. Flick
April 1989

NOTICE

To the best of our knowledge the information in this publication is accurate; however, the Publisher does not assume any responsibility or liability for the accuracy or completeness of, or consequences arising from, such information. This industrial guide does not purport to contain detailed user instructions, and by its range and scope could not possibly do so. Mention of trade names or commercial products does not constitute endorsement or recommendation for use by the Author or Publisher.

Some advanced cleaning products could be toxic if used improperly, and therefore due caution should always be exercised in the use of these materials. Final determination of the suitability and reliability of any information or product for use contemplated by any user, and the manner of that use, is the sole responsibility of the user. We strongly recommend that users seek and adhere to a manufacturer's or supplier's current instructions for handling each material they use.

The Author and Publisher have used their best efforts to include only the most recent data available. The reader is cautioned to consult the supplier in case of questions regarding current availability.

Contents and Subject Index

SECTION I
HOUSEHOLD/INDUSTRIAL CLEANERS

1. BATHROOM CLEANERS 2
   Acid Tile and Bath Cleaner (Pearlized). 3
   Tile Cleaners. 3
   Liquid Acid Toilet Bowl Cleaner 5
   Liquid Acid Toilet Bowl Cleaner (Disinfectant). 6
   Liquid Acid-Type Toilet Bowl Cleaner. 6
   Pearlescent Toilet Bowl Cleaner 7
   Toilet Bowl Cleaners 7
   Toilet Bowl Cleaner-Liquid 8
   Toilet Bowl Cleaner-Solid 9
   Toilet Bowl Cleaners, Solid Automatic 9

2. DISHWASHING DETERGENTS 11
   Dishwash Liquid Concentrate 12
   Dishwash Liquid from Concentrate—Premium Quality 12
   Dishwash Liquid from Concentrate—Good Quality. 13
   Dishwash Liquid from Concentrate—Economy 13
   Dishwash Liquid from Concentrate—Generic. 13
   Dishwash Liquid—Premium Quality. 14
   Dishwash Liquid—Good Quality. 14
   Dishwash Liquid—Economy. 15
   Dishwash Liquid—Generic 15
   Hand Dishwash Liquid—High Quality 16
   Hand Dishwash Liquid—Good Quality. 16
   Hand Dishwash Liquid—Economy (LAS/Ethoxysulfate/Ethoxylate) . . . 17
   Liquid Dishwash (SLS/Sulfonic Acid-TEA/Surfactant). 17

High Active Dishwashing Detergent (30%) 18
High Active Dishwashing Detergent (30%) with Skin Conditioner 19
Light Duty Liquid Dishwash 20
Light Duty Liquid Formulation Medium Cost Hand Dishwash 20
Liquid Dishwashing Compound 21
Liquid Dishwashing Detergents 21
Liquid Dishwashing Detergent (Premium) 23
Liquid Dishwashing Detergent (Medium) 23
Hand Dishwashing Detergent 24
Dishwashing Detergent (24%) 25
Pearlescent Dishwashing Detergent 25
Dishwashing Detergents Liquid, Hand 26
Generic Dishwashing Detergent (14%) 27
Medium Active Dishwashing Detergent (25%) 28
Concentrated Liquid Machine Dishwashing Compound 29
Institutional Liquid Machine Dishwashing Compound 29
Medium Duty Liquid Machine Spray Wash 29
Automatic Dishwashing Detergent (Powder) 30
Mechanical Dishwashing Detergents 30
Dishwashing Detergents, Machine 32
Dishwashing Detergent, Machine (Phosphates/Silicate/Surfactant) 34
Dishwashing Detergent, Machine/Biodegradable 35
Machine Dishwashing Detergent 35
Dishwashing Detergent, Machine 36
Dishwashing Detergent, Machine/Biodegradable 36
Industrial Dishwash Formulation: Regular Duty, Good Quality,
   Mild Liquid 37
Industrial Dishwash Formulation: Good Quality Pot and Pan
   Cleaner 37
Industrial Dishwash Formulation: Alkaline Pot and Pan Cleaner 37
Liquid Machine Dishwashing Concentrate 38
Machine Dishwashing Detergent 38
Automatic Dishwasher Detergent 38
Machine Dishwashing Formulation 39
Machine Dishwashing Powder 39

3. DISINFECTANTS 40
Disinfectant Cleaner 41
General Disinfectant Solution 41
General Disinfectant 41
Disinfectant Cleaners 42
Hospital Disinfectant 43
Sanitizer—Liquid, Acid 43
Sanitizer (Also for Spray)—Liquid, Clear 43
Sanitizer: Liquid, Cloudy, Contains Chlorine 44
Sanitizing Cleaner (Liquid, 12.5% A.M.) 44

4. FLOOR CLEANERS AND WAX STRIPPERS 45
Floor Cleaners 46

Floor Cleaner (Garage)..50
Floor Cleaner—Low Foam Machine Applied....................51
Floor Cleaner, Low-Foam Machine Scrubbers.................51
Floor Cleaner with FBR.......................................53
Floor Cleaner—Garage Floor.................................53
Floor Cleaner/Wax Strippers.................................54
Floor Wax Stripper..56
Floor Cleaner/Wax Stripper—Powder..........................57
Floor Cleaner/Wax Strippers.................................57
Wax Stripper and Cleaner....................................58
Liquid Non-Phosphate Floor Cleaner and Wax Stripper........59
Floor Wax Stripper..59
Experimental Stripper.......................................60
Heavy Duty Wax Stripper.....................................60
Floor Finish Stripper.......................................61
Foaming Wax Stripper..61
Wax Stripper..61
Low Foaming Wax Strippers...................................62
Wax Strippers...63
Waxed Floor Cleaner...64
Heavy Duty Floor Finish Stripper............................64

5. GENERAL PURPOSE CLEANERS

5. GENERAL PURPOSE CLEANERS 65
Abrasive Cleaners, Liquid...................................66
Alkaline All Purpose Industrial Cleaner.....................68
All-Surface Household Cleaner...............................68
All-Purpose Cleaners..69
All-Purpose Cleaner (Aerosol)...............................72
All-Purpose Cleaner with FBR................................73
All-Purpose (Wall/Tile/Floor Cleaner).......................73
All-Purpose Janitorial Cleaning Concentrate.................73
All-Purpose Concentrate Cleaners............................74
All-Purpose Cleaner (Liquid Concentrate)....................75
All-Purpose Liquid Concentrate Cleaner......................75
All-Purpose Heavy Duty Cleaner..............................76
All-Purpose Liquid Industrial Cleaner.......................76
All-Purpose Liquid Detergent................................76
All-Purpose Household Detergent.............................77
All-Purpose Industrial Cleaner..............................77
All-Purpose Light Duty Liquid Detergent.....................78
All-Purpose Cleaner, Light Duty.............................78
All-Purpose Cleaner...78
All-Purpose Liquid Concentrate Cleaner......................79
All-Purpose Cleaner, Liquid.................................79
All-Purpose Cleaner...79
All-Purpose Liquid Detergents...............................80
All-Purpose Liquid Pine Oil Detergent.......................81
All-Purpose Spray Cleaner...................................82

All-Purpose Cleaner with Alkanolamides 82
Caustic Cleaner. 83
Chlorine Cleaner (Kitchen and Bathroom) (Powder). 83
Food Industry Cleaner—General Use Liquid 83
Food Industry Cleaner—Powder. 84
General Cleaner—Alkaline Cleaner 84
General Cleaner—Household 84
General Purpose Cleaner—Household. 85
General Purpose Cleaner—Industrial 85
General Purpose Cleaner. 85
General Purpose Spray and Wipe Cleaners 86
General Purpose Cleaner. 88
Heavy Duty Alkaline Cleaner. 88
Heavy Duty Cleaner. 88
Heavy Duty, Good Quality, All Purpose Spray, Wall/Tile/Floor
  Cleaner 89
Heavy Duty Cleaner. 89
Heavy Duty All Purpose Steam Cleaner 90
Industrial Cleaner 90
Light Duty Cleaner 90
Light Duty, All Purpose Wall/Tile/Floor Cleaner 91
Liquid Concentrate 91
Liquid High Pressure Cleaner (Concentrate) 91
Limonene Based All Purpose Cleaner, Clear Liquid 92
Limonene Based Heavy Duty Cleaner/Degreaser 92
Limonene Based Household Cleaner 93
Limonene Based Household and Industrial Cleaner 94
d-Limonene Spray Cleaner 94
Liquid Cleanser. 95
Liquid General Purpose Alkaline Steam Cleaner—Heavy Duty 95
Multi-Purpose Cleaner—Stripper—Cleaner—Degreaser 95
Multi-Purpose Cleaner Based on Limonene. 96
General Purpose Cleaner. 96

6. LAUNDRY PRODUCTS. 97
Cold Water Detergent (Wool and Delicate Fabrics). 98
Commercial Laundry Liquid 98
Detergents for Fine Fabrics (Clear, Liquid, Without Phosphate) 99
Detergent for Fine Fabrics (Liquid, with Phosphate) 100
Fine Fabric Wash Detergent. 100
Detergent Paste with Phosphate 101
Detergent Paste Without Phosphate. 101
Detergent—Softeners 102
Dry-Blended Laundry Powder Containing Phosphate—Premium
  Quality 104
Dry-Blended Laundry Powder Containing Phosphate—High Quality . . 104
Dry-Blended Laundry Powder Containing Phosphate—Premium
  Quality 105

Dry-Blended Laundry Powder Containing Phosphate—Good Quality . . 105
Dry-Blended Laundry Powder Containing Phosphate—Good
Quality 106
Dry-Blended Laundry Powder Containing Phosphate—Regular
Quality 106
Hand-Wash Laundry Liquid—Moderate Foam 107
Hand-Wash Laundry Liquid—High Foam, Extra Mild 107
Heavy Duty Commercial Laundry Detergent. 108
Industrial Laundry Detergent. 108
Heavy Duty Liquid Detergent (Transparent, Without Builder) 109
Heavy Duty Powder. 109
Heavy Duty Liquid Laundry Detergent "HDL". 110
Heavy Duty Liquid Without Phosphate 110
Heavy Duty Liquid Laundry Detergents (36%) with Fabric
Softener. 111
Heavy Duty Liquid with Phosphate. 112
Heavy Duty Liquid Detergent 112
Heavy-Duty Phosphate Free Laundry Powders 113
Heavy-Duty Powdered Laundry Detergent 113
Industrial Laundry Detergents 114
Laundry Detergent—Commercial: High Quality Built Laundry
Liquid 115
Laundry Detergents. 115
Laundry Detergents—Commercial—Unbuilt Liquids 116
Laundry Liquid Concentrate 117
Laundry Liquid from Concentrate—Good Quality 117
Laundry Liquid from Concentrate—Store Brand 118
Laundry Liquid from Concentrate—Economy. 118
Laundry Liquid from Concentrate—Generic 118
Laundry Liquid with Enzymes—Premium Quality 119
Laundry Liquid with Enzymes—Good Quality 119
Laundry Liquids with Fabric Softener—Premium Quality 120
Laundry Liquid with Fabric Softener—Good Quality 120
Laundry Liquid with Fabric Softener and Enzymes 121
Laundry Paste. 122
Light Duty Liquid Detergent. 122
Concentrated Liquid Laundry Detergent 122
Laundry Powders—Commercial 123
Light Duty Liquids 124
Liquid Detergents—Softener 125
Liquid Detergents 128
Liquid Laundry Detergent. 133
Laundry Detergent, Liquid-Emulsion Type. 134
Laundry Detergent, Liquid—Slurry Type 134
Laundry Detergent, Liquid 135
Laundry Detergent Liquid, Heavy Duty. 136
Laundry Detergent Powder 136
Laundry Detergent Powders (Low-Foam). 137

Laundry Detergent Powder—Low Phosphate. 138
Laundry Detergent Powder—No Phosphate. 138
Laundry Pre-Spotters (Water Based) 139
Liquid Laundry Detergents 141
Heavy-Duty Liquid Laundry Detergent 145
Non-Phosphate Dry-Blended Laundry Powder—Premium Quality 146
Non-Phosphate Dry-Blended Laundry Powder—Good Quality 146
Non-Phosphate Dry-Blended Laundry Powder—Premium Quality 147
Non-Phosphate Dry-Blended Laundry Powder—Good Quality 147
Non-Phosphate Dry-Blended Laundry Powder—Economy 147
Powdered Laundry Detergent (Phosphate Type) 148
Powdered Laundry Detergent (Nonphosphate Type) 148
Regular Quality Built Laundry Liquids 149
Two-Step Built Laundry Liquid Formulations. 151
Unbuilt Laundry Liquid—Super Premium Quality 152
Unbuilt Laundry Liquids—Premium Quality. 152
Unbuilt Laundry Liquid—Regular Quality 153
One-Half Cup Built Laundry Liquid—High Quality. 154
One-Half Cup Built Laundry Liquid—High Quality, High Foam 155
Laundry Prespotter 156
Laundry Prespotter—Aerosol. 156
Aerosol-Type Prespotter—Solvent-Based—Premium Quality 157
Aerosol-Type Prespotters—Solvent-Based—High Quality 157
Laundry Prespotters—Pump Spray-Type—Premium Quality 158
Pump Spray Prespotter—Water-Based—Good Quality 159
Pump Spray Prespotter—Water-Based—Regular Quality. 159
Pump Spray Prespotter—Water-Based—Economy 159

7. METAL CLEANERS 160
Metal Cleaners, Acid 161
Metal Cleaners—Brightener for Aluminum 162
Metal Cleaners, Heavy Duty. 164
Metal Cleaner, Heavy Duty Low-Foam 165
Metal Cleaner, Heavy Duty Low-Foam, Low Temperature. 166
Metal Cleaner, Light Duty. 166
Metal Cleaner, Light Duty for Oil Rigs. 166
Metal Cleaners, Low-Foam Liquid 167
Metal Cleaner, Medium Duty 168
Metal Cleaners, Soak Tank 168
Metal Cleaner—Spray on Type 169
Metal Cleaner for Copper and Brass (Metal Polish). 169
Metal Cleaner, Solvent-Emulsifier. 170
Acid Metal Cleaners. 170
Aircraft Cleaners. 171
Alkaline Metal Degreasing Baths. 172
Alkaline Soak Tank Cleaner for Aluminum. 172
Alkaline Soak Tank Cleaner for Brass 173
Aluminum Brightener/Cleaner 173

Aluminum Cleaner and Brightener (Non-Hydrofluoric Type) 173
Alkaline Metal Cleaner—High Quality 174
Alkaline Metal Cleaner—Good Quality 174
Aluminum Cleaner—Acid Aluminum Brightener 175
Aluminum Cleaner—Alkaline Aluminum Cleaner with Phosphate 175
Aluminum Brightener 176
Aluminum Cleaners 176
Caustic-Gluconate Soak Formulation 178
Cold Metal Cleaners 178
Copper Cleaner (Paste) 179
Metal Cleaner (Soak Cleaner) 179
Heavy-Duty Metal Cleaner 179
Metal Cleaner—Alkaline Wash Powder—with Phosphate 180
Metal Cleaner—Alkaline Wash Powder—Non-Phosphate Powder 180
Metal Cleaner—Paint Stripper 180
Metal Cleaning—Paint Removal 181
Metal Cleaning—Rust Removal 181
Metal Cleaning—Alkaline 181
Metal Cleaning—Acid 181
Metal Cleaner, Powder, for Spray Type 182
Soak Tank Metal Cleaner, Powder, for Aluminum 182
Rust Remover (Dip Part Cleaners) 183
Silver Cleaning Bath 183
Soak Tank Metal Cleaner 183
Soak Tank Metal Cleaner, Powder, for Brass 184
Soak Tank Metal Cleaner, Powder, for Zinc 184
Soak Tank Metal Cleaner, Powder, for Magnesium 185
Soak Tank Metal Cleaner, Powder, for Copper 185
Spray Metal Cleaners 186
Stem Metal Cleaners 186
High Quality Metal Deoiling Liquid Concentrates (Oil Spill/Rig
   Cleaners) 187

8. OVEN CLEANERS 188
Aerosol Oven Cleaner 189
Oven Cleaners 189
Oven Cleaner, Liquid Type 190
Oven Cleaner, Foam Spray Type 191
Oven Cleaner 191
Caustic Cleaners (Oven Cleaner or Paint Stripper) 192

9. RINSE ADDITIVES AND AIDS 194
Rinse Additives Biodegradable, Machine Dishwashing 195
Rinse Additives, Machine Dishwashing 196
Rinse Aid Concentrates 199
Dishwasher Rinse Aid 199
Rinse Aid—Economical Type 200
Rinse Aid—Low Density Type 200

Rinse Aid—High Density Type. 201
Dishwasher Rinse Aids. 201
Rinse Aid for Dishwash 202
Rinse Aid with Makon NF-5 202
Rinse Aids for High Temperature Machines 203
Rinse Aids for Low Temperature Machines. 205
Rinse Aids for Low Actives Formulations (10 wt % Surfactants) . . . 208
Rinse Aids for Hard Water Formulations 210

10. RUG, CARPET AND UPHOLSTERY CLEANERS AND SHAMPOOS. . . 212
Antistatic Carpet Shampoo. 213
Carpet Cleaner 214
Carpet Cleaner (Foam Type) 214
Carpet Cleaner (Scrub Type) 214
Carpet Extraction Cleaner. 215
Carpet Extractor Cleaner 215
Carpet Steam Cleaner. 215
Carpet Shampoo. 216
Carpet Shampoo, Aerosol 217
Carpet Shampoos, Scrubbing Machine. 217
Carpet Shampoos, Steam Cleaner 218
Carpet Shampoo Treatment, Soil Retardant and Antistat Cleaning
   Technique—Pump Spray. 220
Carpet Shampoo Treatment, Soil Retardant and Antistat Cleaning
   Technique—Aerosol 220
Liquid Spray Vacuum Carpet Cleaner 221
Powdered Spray Vacuum Carpet Cleaner 221
Premium Medium Duty Steam Extraction Carpet Cleaner 222
Premium Heavy Duty Steam Extraction Carpet Cleaner with Butyl. . . 222
Heavy Duty Steam Extraction Carpet Cleaner. 222
Rug Cleaners 223
Aerosol Rug Shampoo with Anticorrosive 224
Rug Shampoo with Polymer 224
Basic Rug Shampoo Concentrate 225
Rug and Upholstery Shampoo—Low-Foaming Liquid. 225
Rug and Upholstery Shampoo—High-Foaming Liquid. 225
Rug Shampoos 226
Rug Shampoo, Liquid Type. 227
Rug Steam Cleaner 228
Rug Steam Cleaner—Non-Phosphate Powder. 228
Rug Steam Cleaner—Powder with Phosphate. 228
Rug and Upholstery Cleaner—"Dry Brittle" Liquid Cleaner—High
   Quality 229
Rug and Upholstery Cleaner—"Dry Brittle" Liquid Cleaner—Good
   Quality 229
Rug and Upholstery Cleaner—"Dry Brittle" Liquid Cleaner—
   Economy. 230
Liquid Steam Cleaner for Carpets for Use in Hard Water Systems . . . 230

11. WALL AND HARD SURFACE CLEANERS 231
    All Purpose Hard Surface Cleaning Concentrates 232
    Hard Surface Cleaners 232
    Hard Surface Cleaner—All Purpose Creamy Scouring Cleanser 234
    Hard Surface Cleaner: All-Purpose Spray. 235
    Hard Surface Cleaner: Liquid Concentrate. 235
    Hard Surface Cleaner: Liquid Disinfectant. 235
    Hard Surface Cleaner: Floor Cleaner/Wax Stripper 235
    Hard Surface Cleaner, Liquid, All Surface Bathroom Type 236
    Hard Surface Cleaner, All Surface Bathroom—Acid Type 236
    Hard Surface Cleaner, Degreasing Type 237
    Hard Surface Cleaners, All Purpose Type 237
    Hard Surface Cleaner 239
    Hard Surface Cleaner, General Purpose Type 240
    Hard Surface Cleaner, Heavy Duty Type 240
    Hard Surface Cleaner, Pine Oil Type—Economical 241
    Hard Surface Cleaner, Solvent Type 241
    Hard Surface Cleaners—All Purpose. 242
    Hard Surface Cleaners—All Purpose Non-Phosphate Premium
    Quality Liquid Concentrates 243
    Hard Surface Cleaner Liquid Concentrates, High Quality with
    Phosphate 244
    Hard Surface Cleaners—All Purpose Hard Surface Cleaner Liquid
    Concentrates. 245
    Hard Surface Cleaners—All Purpose. 246
    Hard Surface Cleaners—All Purpose Pine Oil Cleaners 247
    Hard Surface Cleaners—All Purpose. 248
    Hard Surface Sanitizer and Cleaner Disinfectant 249
    Hard Surface Cleaner/Fabric Pretreat 249
    Hard Surface Spray Cleaner. 250
    Hard Surface and Floor Cleaner. 250
    Liquid Hard Surface Cleaner, High Sudsing—Unbuilt 251
    Liquid Hard Surface Cleaner, Dilutable, Phosphate Built. 251
    Liquid Hard Surface Cleaner Concentrates 252
    Wall Cleaner. 252
    Wall Cleaner, Light Duty. 252
    Wall Cleaner, Spray 252

12. WINDOW AND GLASS CLEANERS. 253
    Acid Glass Cleaner. 254
    Multi-Feature Glass Cleaner. 254
    Ammoniacal WINDEX Type 255
    Window/Glass Cleaner (Good Quality). 255
    Glass Cleaner 256
    Glass Cleaners (All Purpose) 256
    Glass Cleaner, Ready to Use 258
    Glass Cleaners. 258
    Glass Cleaner Concentrate. 259

Industrial Glass Cleaner 260
Liquid Glass Cleaner 260
Window/Glass Cleaner (Good Quality with Alcohol) 261
Window and Glass Cleaners 261
Window Cleaners 262
Window Cleaner, Aerosol (Foam-Type) 263
Window Cleaner, High Alcohol Type 264
Window Cleaner, Concentrate 264
Window Cleaners, Spray 265

13. MISCELLANEOUS CLEANERS 266
Abrasive Cleaner—High Viscosity 267
All-Surface Steam Cleaner 267
Alkaline Cleaner 268
Asphalt Release Agents 268
Barbecue Cleaner: High Alkaline 269
Bottle Cleaner 269
Bottle Wash Concentrate 269
Bilge Cleaner—Solvent Emulsifier 270
Bilge Cleaners 270
Bottle-Wash Compound 271
Bottle Washes, Machine 271
Bowl Cleaner 272
Butyl Cleaners (Heavy Duty Degreaser) 272
Butyl Cleaner and Degreaser 273
Carbon Cleaner 274
Cosmoline Remover (Heavy Oil or Greasy-Wax Remover) 274
Chain Lubricant 275
CIP Cleaners 275
Coffee and Tea Machine Cleaner (Concentrate) 275
Concrete Cleaners 276
Cutting and Grinding Fluid (Low Foam) 276
Dairy Milkstone Remover 277
Dairy Pipeline Cleaner (Low-Foam) 277
Dairy Line Cleaners 278
Dairy Pipeline Cleaner 278
Dairy Cleaners 279
Degreasers, Degreaser Concentrates 280
Degreasers, Solvent Degreasers, Flush-Off Type 282
Powder Degreasers 282
Degreaser 283
Spray Degreaser 283
Heavy Duty Spray Degreaser 283
High Foam Degreaser 284
Non-Butyl Degreaser 284
Non-Phosphate Non-Butyl Degreaser 284
Industrial Strength Non-Butyl Degreaser 285
Water-Based Solvent Degreaser 285

Drain Cleaners, Liquid 286
Drain Cleaner, Liquid Type 286
Dry Cleaning Solutions 287
Dry Cleaning Compound 287
Dry Cleaning Fluid 287
Dry Cleaning Formulas 288
Emulsifiable Solvent Cleaners (Garage Floor Cleaner) 289
Foaming Butyl Cleaner 289
Foam Marker Concentrates 290
Food Processing Equipment Cleaner, Spray 290
Gel Rust Remover 291
Graffiti Removers 291
Glass/Bottle Liquid Cleaner Compound 292
Grill Cleaner 292
Heavy Duty Cleaner 292
Heavy-Duty Cleaner—Dry 293
Heavy Duty Cleaner and Degreaser 293
Heavy Duty Concrete Cleaner 293
Heavy Duty Household Cleaner 294
Heavy Duty Liquid Steam Cleaner 294
Heavy Duty Paint Stripper 295
High Alkaline Cleaners 295
High Pressure Cleaner, Liquid, Clear 296
High Pressure Cleaner, Acid 296
High Pressure Steam Cleaner 296
Hot Plate/Grill Cleaner 297
Industrial Cleaner, Alkaline 297
Liquid High Pressure Cleaner Concentrate 298
Liquid High Pressure Concentrate 298
Liquid Acid Dairy Sanitizer and Cleaner 298
Liquid Rust Remover 299
Liquid Caustic Cleaner 299
Low Foam Spray Washing Compound 299
Lipstick Stain Remover 299
Leather, Vinyl, Plastic Cleaner 299
Leather Cleaner 300
Medium Duty Steam Cleaner 300
Oil Field Apparatus Cleaner 300
Medium Duty Non-Caustic Steam Cleaner 301
Paint Brush Cleaner 301
Produce Peeling Formula 301
Porcelain Cleaner 302
Heavy Duty Pot and Pan Cleaner 302
Powdered Caustic Cleaner 303
Rust Remover—Liquid-Acid 303
Rust Remover—Liquid-Alkaline 303
Solvent Emulsion, Sludge and Carbon Cleaner 304
Solvent Cleaner 304

Spotting Liquid for Dry Cleaning 304
Spot Cleaner. 305
Synthetic Cleaner 305
Steam Cleaner, Liquid 306
Steam Cleaner Powder 306
Steam Cleaning Compound 306
Steam Cleaners. 307
Steam Cleaning Compounds 309
Textile Softener 309
Tank Cleaners. 310

SECTION II
AUTOMOTIVE CLEANERS

14. CAR AND TRUCK WASHES 312
Auto Shampoo. 313
Auto Shampoo (Non-Streaking) 313
Automobile Foaming Spray. 313
Boat Wash and Car Cleaner 313
Car Wash Detergent Powders 314
Car Wash, Liquids 315
Car Wash, Thickened Liquid 316
Liquid Car Washes. 316
Liquid Car Wash Concentrate. 316
Car Wash, Powdered Type. 317
Car Wash, Liquid Type 317
Car Wash Liquid Concentrates 318
Car Wash Powders 319
Car Wash Concentrates with Alcohol Ethoxylate. 320
Medium-Cost Car Wash. 320
Car Wash Detergent 321
Hand Car Wash 321
Car Shampoos. 322
Car Cleaner 322
Car Spray Rinses. 323
Automatic Car Wash 323
Manual Car Wash—Home Use. 323
Car Spray Wash. 324
Wand-Type Car Wash 324
Medium Duty Truck and Rig Wash, Powder 325
Heavy Duty Truck Cleaner 325
Truck Washes. 326
Liquid Truck Washes 327
Truck—Car Wash. 328
Liquid Spray Cleaner for Aluminum Trucks 328
Vehicle Washes 328

15. WHITEWALL TIRE CLEANERS 329
    Automobile Whitewall Tire Cleaner 330
    Foamy Whitewall Tire Cleaner, Liquid 330
    ''White Lightning'' Tire Cleaner 331
    White Wall Tire Bleach 331
    White Sidewall Tire Cleaner 331
    White Wall Tire Cleaners 332

16. MISCELLANEOUS CLEANERS 333
    Acidic Cleaner for Aluminum Trucks 334
    Carburetor Cleaner 334
    Automobile Windshield Cleaners 335
    Engine Degreaser Concentrate 336
    Exterior Railcar Cleaner 336
    Powdered Hubcap Cleaner 336
    Solvent Emulsion Cleaner 337
    Vinyl Top Cleaner 337
    Windshield Washer Cleaners 338
    Windshield Washer Formulations 339
    Windshield Wash 340
    Windshield Washer Formulation 340

SECTION III: TRADEMARKED RAW MATERIALS 341

SECTION IV: SUPPLIERS' ADDRESSES 369

Section I
Household/Industrial Cleaners

1. Bathroom Cleaners

ACID TILE & BATH CLEANER(PEARLIZED)

RAW MATERIALS	% By Weight
VARION CADG-HS	10.0
Citric Acid	5.0
REWOPOL PCK2000(cold-mix pearl)	4.0
REWOPOL PEG 6000 DS (PEG 150 Distearate)	3.0
VAROX 1770	3.0
Water	75.0

Mixing Procedure:
   Add PEG 6000 DS to 60C water to disperse.
   Add the VARION CADG-HS while hot.
   Cool to 30C and then add the Citric Acid, PGK2000 and VAROX
      1770

SOURCE: Sherex Chemical Co.: Industrial Formulation 51/01.7

TILE CLEANER

RAW MATERIALS	% By Weight
Water	72.5
Phosphoric Acid (85%)	12.0
Hydroxyacetic Acid (50%)	10.0
MAZER MAZON 41	4.0
Versene	0.5
Kelzan Gum Thickener	1.0

SOURCE: Mazer Chemicals, Inc.: Household/Industrial T-20B:
         Formula 12

TILE CLEANER

RAW MATERIALS	% By Weight
Water	60.7
PLURAFAC D-25 surfactant	5
Colloidal magnesium aluminum silicate	4.3
Calcium carbonate	30

Use as is

SOURCE: BASF CORP.: Cleaning Formulary: Formulation #3425

TILE CLEANER

RAW MATERIALS	% By Weight
Water	77
Sodium xylene sulfonate (40%)	6
Ethylene glycol butyl ether	4
Linear alkyl aryl sulfonate (60%)	2
PLURAFAC D-25 surfactant	2
Sodium metasilicate pentahydrate	3
Tetrapotassium pyrophosphate	6

Use as is

SOURCE: BASF CORP.: Cleaning Formulary: Formulation #3246

TILE CLEANER

RAW MATERIALS	% By Weight
Water	96
Sodium Hypochlorite	2
Sodium Hydroxide	1
AVANEL S-70	1

Mixing Procedure
    Charge vessel with ingredients in the order listed with agitation

PROPERTIES:
    This is a liquid tile cleaner suitable for a pump spray application. It has the advantage of leaving a streak-free shine with no residue.

SOURCE: Mazer Chemicals, Inc.: AVANEL S Formula: Tile Cleaner
        JM-03

LIQUID ACID TOILET BOWL CLEANER

RAW MATERIALS	% By Weight
NEODOL 25-12	3.0
Bardac 22 (50%)	5.0
Hydrochloric acid (35%)	20.0
Cocobetaine (30%)	3.
Water, dye	to 100%

PROPERTIES:
    Viscosity, 73F, cps   7
    Phase coalescence temp., F   >176
    pH   1.0

SOURCE: Shell Chemical Co.: NEODOL Formulary: Suggested
        Formulation

LIQUID ACID TOILET BOWL CLEANER(DISINFECTANT)(HIGH QUALITY)

RAW MATERIALS	% By Weight
NEODOL 25-12	3.0
Bardac 22 (50%)	5.0
Hydrochloric acid (35%)	50.0
Cocobetaine (30%)	3.0
Water, dye	to 100%

Properties:
Viscosity, 73F, cps	11
Phase coalescence temp., F	>176
pH	1.0

SOURCE: Shell Chemical Co.: NEODOL Formulary: Suggested
       Formulation

LIQUID ACID-TYPE TOILET BOWL CLEANER

RAW MATERIALS	% By Weight
SURFONIC N-60	10
Kerosene	17
Phosphoric Acid	13
Water	60

Mix water, SURFONIC N-60 and acid, then add kerosene with rapid agitation; dilute 1 to 100 with water. Use at 180 to 190F.

SOURCE: Texaco Chemical Co.: Suggested Formulation

PEARLESCENT TOILET BOWL CLEANER(2878-029)

RAW MATERIALS   % By Weight

A
Water   34.0
TRYFAC 5552 Phosphate Ester   1.0
Triethanolamine (TEA)   1.0
EMID 6500 Coconut Monoethanolamide   4.0

B
Water   52.0
Tetrasodium EDTA (40%)   5.0
Sodium metasilicate pentahydrate   2.0
Ethylene glycol n-butyl ether   1.0

Blending Procedure:
    Part A: Add the water to the first blending tank and heat
to 70C. While mixing, add the other ingredients listed in Part A.
Mix until uniform and then cool to room temperature.
    Part B: Add the water to the second blending tank. While
mixing, add the other ingredients listed in Part B. Mix until
uniform.
    Add Part B to Part A and mix until uniform.

SOURCE: Emery Chemicals: Specialty Chemicals Formulary:
        Formulation 2878-029

TOILET BOWL CLEANER(2878-029)

RAW MATERIALS   % By Weight

Tetrasodium EDTA (40%)   5.0
TRYCOL 5941 POE (9) Tridecyl Alcohol   0.5
TRYFAC 5559 Phosphate Ester   0.3
Sodium carbonate   1.0
Ethylene glycol n-butyl ether   5.0
Dye, fragrance, etc.   q.s.
Water   to 100

Blending Procedure:
    Add the water to the blending tank. While mixing, add the
remaining ingredients in the order listed. Mix until uniform.

SOURCE: Emery Chemicals: Specialty Chemicals Formulary:
        Formulation 2878-029

<u>TOILET BOWL CLEANER S-34</u>

RAW MATERIALS   % By Weight

Deionized Water   87.5
AVANEL S-90   5.0
37% HCl   7.5

Mixing Procedure:
   Cnarge vessel with ingredients in the order listed using
moderate agitation

   This is a liquid toilet bowl cleaner, comparable in effi-
ciency to the solid cleaners containing sodium bisulfide. It
has the added advantage of superior detergency, with the con-
venience of a liquid. It can be applied from a plastic squeeze
bottle. The other AVANELS could be used in this formula as
desired.

SOURCE: Mazer Chemicals, Inc.: AVANEL S Formula: S-34

<u>TOILET BOWL CLEANER-LIQUID</u>

RAW MATERIALS   % By Weight

NEODOL 25-12   5.0
Urea   10.0
EDTA (a)   0.5
Water, dye, perfume   to 100%

   (a) Ethylenediamine tetraacetic acid, tetrasodium salt (100%
       basis).

Properties:
   Viscosity, 73F, cps   5
   Phase coalescence temp., F   >176
   pH   10.6

SOURCE: Shell Chemical Co.: NEODOL Formulary: Suggested
        Formulation

TOILET BOWL CLEANER-SOLID

RAW MATERIALS	% By Weight
NEODOL 25-12	10.0
PEG (1400) (b)	10.0
Urea	80.0

(b) Polyethylene glycol with molecular weight about 1400.

SOURCE: Shell Chemical Co.: NEODOL Formulary: Suggested
          Formulation

TOILET BOWL CLEANER, SOLID AUTOMATIC

RAW MATERIALS	% By Weight
TRITON N-998 Surfactant (a)	25.0
Urea	10.0
Sodium Sulfate, Anhydrous	65.0
	100.0

   Biocides, water soluble dyes, perfumes or silicone defoamers
can be added.

   (a) TRITON N-998 Surfactant-100%

Mixing Instructions:
   Stir and heat a mixture of TRITON N-998 Surfactant and urea
until it becomes clear. Add sodium sulfate and mix thoroughly.
Continue mixing while cooling. Discharge to a suitable container.
The product is a semi-solid which compacts on standing.

Lit. Ref.: CS-435

SOURCE: Rohm and Haas Co.: Specialty Chemicals: Detergent Form-
          ulations for Industrial and Institutional Industry

TOILET BOWL CLEANER, SOLID AUTOMATIC

RAW MATERIALS	% By Weight
TRITON N-998 Surfactant (a)	35.0
Urea	5.0
Sodium Sulfate, Anhydrous	60.0
	100.0

Biocides, water soluble dyes, perfumes or silicone defoamers can be added.

(a) TRITON N-998 Surfactant-70%

Mixing Instructions:
Stir and heat a mixture of TRITON N-998 Surfactant and urea until it becomes clear. Add sodium sulfate and mix thoroughly. Continue mixing while cooling. Discharge to a suitable container. The product is a semi-solid which compacts on standing.

Lit. Ref.: CS-435

SOURCE: Rohm and Haas Co.: Specialty Chemicals: Detergent
        Formulations for Industrial and Institutional Industry:
        Suggested Formulation A

TOILET BOWL CLEANER, SOLID AUTOMATIC

RAW MATERIALS	% By Weight
TRITON N-998 Surfactant (100%)	18.0
Urea	82.0
	100.0

Biocides, water soluble dyes, perfumes or silicone defoamers can be added.

Mixing Instructions:
Heat TRITON N-998 Surfactant in a stirred reactor until it is molten. Add urea (melting point 133C.) while increasing the temperature. Continue stirring until the mixture is a uniform, clear liquid. Package while hot. Upon cooling, the product will solidify to a hard mass.

Lit. Ref.: CS-435

SOURCE: Rohm and Haas Co.: Specialty Chemicals: Detergent
        Formulations for Industrial and Institutional Industry

2. Dishwashing Detergents

DISHWASH LIQUID CONCENTRATE

RAW MATERIALS   % By Weight

Linear alkylbenzene sulfonate(100%) (a)	27.0
NEODOL 25-3A(60%)	22.5
FADEA (b)	4.5
Preservative(s)	0.1
Water	to 100%

    (a) Dodecylbenzene sulfonic acid (DDBSA) plus an equivalent amount of caustic (NaOH) can be used. pH should be between 5.5 and 7.5 before NEODOL 25-3A is added (adjust with more DDBSA or caustic)
    (b) Fatty acid diethanol amide.

Properties:
    Viscosity, 73F, cps   ~700
    Active matter, %w   45

Blending Procedure:
    Dissolve preservative and alkylbenzene sulfonate in water. Add NEODOL 25-3A slowly to well stirred mixture, then add amide. Best results will be obtained if water is warm (e.g. 120F).

SOURCE: Shell Chemical Co.: The NEODOL Formulary: Suggested
        Formulation

DISHWASH LIQUID FROM CONCENTRATE--PREMIUM QUALITY

RAW MATERIALS   % By Weight

Concentrate	71.1
Ethanol SD-3A	4.0
Water, dye, perfume	24.9

Properties:
    Viscosity, 73F, cps   240
    Clear point, F   32

SOURCE: Shell Chemical Co.: The NEODOL Formulary: Suggested
        Formulation

DISHWASH LIQUID FROM CONCENTRATE--GOOD QUALITY

RAW MATERIALS   % By Weight

Concentrate   55.6
Ethanol SD-3A   2.0
Water, dye, perfume   42.4

Properties:
    Viscosity, 73F, cps   235
    Clear point, F   41
Blending Procedure:
    Dissolve the ethanol or sodium sulfate in the water. Add the
concentrate slowly with stirring. Adjust the pH to 6.5-7.0 using
citric acid.

DISHWASH LIQUID FROM CONCENTRATE--ECONOMY

RAW MATERIALS   % By Weight

Concentrate   44.4
Water, dye, perfume   55.6

Properties:
    Viscosity, 73F, cps   195
    Clear point, F   39
Blending Procedure:
    Dissolve the ethanol or sodium sulfate in the water. Add the
concentrate slowly with stirring. Adjust the pH to 6.5-7.0 using
citric acid.

DISHWASH LIQUID FROM CONCENTRATE--GENERIC

RAW MATERIALS   % By Weight

Concentrate   22.2
Sodium sulfate, anhydrous   3.0
Water, dye, perfume   74.8

Properties:
    Viscosity, 73F, cps   220
    Clear point, F   39
Blending Procedure:
    Dissolve the ethanol or sodium sulfate in the water. Add
the concentrate slowly with stirring. Adjust the pH to 6.5-7.0
using citric acid.

SOURCE: Shell Chemical Co.: The NEODOL Formulary: Suggested
        Formulation

DISHWASH LIQUID--PREMIUM QUALITY

RAW MATERIALS   % By Weight

NEODOL 25-3S(60%)   18.3
C12 LAS (60%)   30.0
FADEA (b)   4.0
Sodium xylene sulfonate (40%)   8.5
Sodium chloride   3.0
Water, dye, perfume, preservatives   to 100%

    (b) Fatty acid diethanol amide.

Properties:
    Viscosity, 73F, cps   270
    Clear point, F   38
    Adjust pH to 6.5-7.0 with citric acid.

SOURCE: Shell Chemical Co.: The NEODOL Formulary: Suggested
        Formulation

DISHWASH LIQUID--GOOD QUALITY

RAW MATERIALS   % By Weight

NEODOL 25-3S (60%)   12.5
C12 LAS (60%)   25.0
FADEA   2.5
Ethanol   3.0
Water, dye, perfume, preservatives   to 100%

    (b) Fatty acid diethanol amide

Properties:
    Viscosity, 73F, cps   134
    Clear point, F   18
    Adjust pH to 6.5-7.0 with citric acid.

SOURCE: Shell Chemical Co.: The NEODOL Formulary: Suggested
        Formulation

DISHWASH LIQUID--ECONOMY

RAW MATERIALS   % By Weight

NEODOL 25-3S (60%)   8.3
C12LAS (60%)   13.5
FADEA (b)   1.9
Sodium xylene sulfonate (40%)   3.0
Sodium sulfate   0.6
Sodium chloride   1.0
Water, dye, perfume, preservatives   to 100%

   (b) Fatty acid diethanol amide

Properties:
   Viscosity, 73F, cps   104
   Clear point, F   18
   Adjust pH to 6.5-7.0 with citric acid.

SOURCE: Shell Chemical Co.: The NEODOL Formulary: Suggested
        Formulation

DISHWASH LIQUID--GENERIC

RAW MATERIALS   % By Weight

NEODOL 25-3S (60%)   5.5
C12 LAS (60%)   9.2
FADEA (b)   1.2
Sodium xylene sulfonate (40%)   1.0
Sodium chloride   2.0
Water, dye, perfume, preservatives   to 100%

   (b) Fatty acid diethanol amide

Properties:
   Viscosity, 73F, cps   209
   Clear point, F   18
   Adjust pH to 6.5-7.0 with citric acid.

SOURCE: Shell Chemical Co.: The NEODOL Formulary: Suggested
        Formulation

HAND DISHWASH LIQUID--HIGH QUALITY

RAW MATERIALS	% by Weight
C12 LAS (60%) (a)	26.0
NEODOL 91-8	4.8
NEODOL 25-3S (60%)	8.0
FADEA (b)	3.2
Ethanol	3.0
Water, dye, perfume, preservatives	to 100%

Properties:

Viscosity, 73F, cps	67
Phase coalescence temp., F	>176
Clear point, F	32
pH	8.6

HAND DISHWASH LIQUID--GOOD QUALITY

RAW MATERIALS	% By Weight
C12 LAS (60%) (a)	16.7
NEODOL 91-8	3.0
NEODOL 25-3S (60%)	5.0
FADEA (b)	2.0
Ammonium chloride	0.2
Water, dye, perfume, preservatives	to 100%

Properties:

Viscosity, 73F, cps	18
Phase coalescence temp., F	>176
Clear point, F	32
pH	8.3

(a) May use the appropriate amount of dodecylbenzene sulfonic
    acid (DDBSA) with an equivalent amount of sodium hydrox-
    ide to neutralize it.
(b) Fatty acid diethanolamide.

Blending Procedure:
    Effective stirring should be maintained during addition of
all ingredients, and each ingredient should be in solution
before the next is added. Best results will be obtained if
water is warm (e.g. 100-120F).
    1. Dissolve the preservative, linear alkylbenzene sodium
       sulfonate (LAS), ethanol (when indicated) and ammonium
       chloride (when indicated) in water.
    2. Add the NEODOL 91-8.
    3. Add the NEODOL 25-3S slowly with efficient stirring.
    4. Add the amide with efficient stirring.
    5. Add perfume and dye as needed to give the desired odor
       and color.

SOURCE: Shell Chemical Co.: The NEODOL Formulary: Suggested
        Formulations

HAND DISHWASH LIQUID-ECONOMY(LAS/ETHOXYSULFATE/ETHOXYLATE)

RAW MATERIALS	% By Weight
C12 LAS (60%) (a)	8.3
NEODOL 91-8	1.5
NEODOL 25-3S (60%)	2.5
FADEA (b)	1.0
Ammonium chloride	0.3
Water, dye, perfume, preservatives	to 100%

(a) May use the appropriate amount of dodecylbenzene sulfonic
    acid (DDBSA) with an equivalent amount of sodium hydrox-
    ide to neutralize it.
(b) Fatty acid diethanolamide.

Blending Procedure:
   Effective stirring should be maintained during addition of
all ingredients, and each ingredient should be in solution
before the next is added. Best results will be obtained if
water is warm (e.g. 100-120F).
   1. Dissolve the preservative, linear alkylbenzene sodium
      sulfonate (LAS), ethanol (when indicated) and ammonium
      chloride (when indicated) in water.
   2. Add the NEODOL 91-8.
   3. Add the NEODOL 25-3S slowly with efficient stirring.
   4. Add the amide with efficient stirring.
   5. Add perfume and dye as needed to give the desired odor
      and color.

SOURCE: Shell Chemical Co.: The NEODOL Formulary: Suggested
        Formulation

LIQUID DISHWASH(SLS/SULFONIC ACID-TEA/SURFACTANT)

RAW MATERIALS	% By Weight
VARSULF S1333	4.0
Sodium Laureth Sulfate (28%)	42.0
Linear Alkyl Benzene Sulfonic Acid/TEA	17.0
Water	qs100

Mixing Procedure:
   Add the ingredients in the order shown.

SOURCE: Sherex Chemical Co.: Industrial Formulation 13:01.2

HIGH ACTIVE DISHWASHING DETERGENT(30%)(2887-051)

RAW MATERIALS	% By Weight
Water	to 100
Caustic soda (50% sodium hydroxide)	3.3
Dodecylbenzene sulfonic acid (DDBSA)	13.0
EMERSAL 6453 Sodium Laureth Sulfate (28%)	20.0
TRYCOL 5967 POE (12) Lauryl Alcohol	2.0
Tetrasodium EDTA	0.5
EMID 6515 Cocamide DEA	3.0
Sodium xylene sulfonate (40%) (SXS)	q.s.
Citric acid (50%) (to pH 6.5-7.5)	q.s.
Dye, fragrance, opacifier and preservative	as desired

Blending Procedure:
   Charge the water to the batching tank and add the raw materials in the order listed. Warm water will facilitate blending of surfactants. The pH of the batch tank should not be higher than 5 after the DDBSA has been added. If not, adjust the pH with sodium hydroxide before proceeding. Add sodium xylene sulfonate to adjust the formula to approximately 150-300 cP (Brookfield Viscometer).

   Adjust the batch to the final pH before the dye (as an aqueous solution) is added. If an opaque, lotion-type product is desired, 0.15% of an opacifier may be added. One part WITCOPAQUE R-25 (Witco) or E-288 (Morton) opacifier must be preblended with three parts water before its addition to the batch tank. The final pH must be adjusted before the opacifier is added.

SOURCE: Emery Chemicals: Specialty Chemicals Formulary:
        Formulation 2887-051

HIGH ACTIVE DISHWASHING DETERGENT(30%) WITH SKIN CONDITIONER
(2878-115)

RAW MATERIALS   % By Weight

Water	to 100
Caustic soda (50% sodium hydroxide)	3.3
Dodecylbenzene sulfonic acid (DDBSA)	13.0
Sodium ethoxylated alcohol sulfate (60% active)	10.0
Sodium xylene sulfonate (40%) (SXS)	5.0
EMID 6515 Cocamide DEA	3.0
ETHOXYLAN 1686 PEG-75 Lanolin	0.5
TRYCOL 5967 POE (12) Lauryl Alcohol	2.0
Citric acid (50%) (to pH to 6.5-7.5)	q.s.
Dye, fragrance, preservative and opacifier	as desired

Blending Procedure:
    Add the water, caustic soda and DDBSA to the batch tank.
The pH of this mixture should be greater than 5.0. A low pH
may cause corrosion of the blending tank and should be adjusted
by the addition of more caustic soda. While mixing, add the
remaining ingredients to the blending tank in the order listed.
Stir until uniform.
Note: If no claims are made regarding skin conditioning effects,
the PEG-75 lanolin can be omitted. This results in improved foam
levels. The viscosity of the finished product can be adjusted by
increasing or decreasing the SXS.

SOURCE: Emery Chemicals: Specialty Chemicals Formulary:
        Formulation 2878-115

LIGHT DUTY LIQUID DISHWASH
Formulation 1

RAW MATERIALS	% By Weight
DIACID H-240	7.5
NaLAS	15.0
Neodol 25-3S	17.2
Monoethanolamine	4.0
Water	q.s.*

Formulation 2

RAW MATERIALS	% By Weight
DIACID H-240	6.0
NaLAS	12.0
Neodol 25-3S	9.5
Diethanolamide	2.5
Water	q.s.*

* q.s.--quantity sufficient to make 100% total

Petroleum-based hydrotropes have been used for many years as stabilizers in liquid dishwash detergents. DIACID surfactants will assure good formulation stability, and unlike other hydrotropes, will contribute to the viscosity. While DIACID is not effective at low pH, when used above pH 7.8, it provides recovery from freezing temperatures without altering foaming performance.

SOURCE: Westvaco Chemical Division: DIACID Surfactants:
          Formulation 1 and 2

LIGHT DUTY LIQUID FORMULATION
MEDIUM COST HAND DISHWASH

RAW MATERIALS	% By Weight
Water, D.I.	41.8
DESONOL SE	18.3
DESONATE 60-S	30.0
Varamide MA-1	4.0
PETRO LBA Liquid	2.8
Sodium Chloride	3.0
Formalin	0.1
Perfume, Dye	q.s.

Blending Procedure: Blend ingredients in the order listed.
    Adjust pH = 6.5-7.0 using Citric Acid.
Typical Properties:
    Viscosity = 190 cps
    Clear Liquid

SOURCE: DeSoto, Inc.: Suggested Formulation

LIQUID DISHWASHING COMPOUND

RAW MATERIALS	% By Weight	CAS Registry Number
Water	70.00	
ESI-TERGE T-60	25.00	27323-41-7
ESI-TERGE S-10	5.00	61789-19-3
	100.00	

Procedure:
    Add in order listed and agitate until uniform.

Specifications:
% Solids	20.0
% Active	20.0
pH	7.0-7.5
Viscosity	Medium

SOURCE: Emulsion Systems Inc.: Technical Service Bulletin
        Code S-10-1

LIQUID DISHWASHING DETERGENT
Manual--High Sudsing
Mixed Surfactant Type

RAW MATERIALS	% By Weight
Water	66.1
Sodium alkylbenzene sulfonate	20.9
IGEPAL CO-710	10.0
GAFAMIDE CDD-518	3.0
	100.0

    Perfume, colorants and opacifiers added, as desired, replacing water.

Manufacturing Procedure:
    1. Dissolve sodium alkylbenzene sulfonate. (Note: Alternatively, alkylbenzene sulfonic acid can be neutralized with sodium hydroxide to yield the same sodium alkylbenzene sulfonate activity.)
    2. Add remaining components in the order listed.

Physical Properties:
pH (as is)	10.6
pH (1%)	8.7
Viscosity	600 cps
Specific Gravity	1.02

SOURCE: GAF Corp.: Formulary: Prototype Formulation GAF 5202

LIQUID DISHWASHING DETERGENT

RAW MATERIALS	% By Weight
Water	62.4
BIO SOFT D-62	25.0
STEOL CS-460	6.5
NINOL 128-EXTRA	6.0
Preservative	0.1

Mixing Procedure:
   Adjust pH to 7.0-7.8
Properties:
   % active: 25.0

Formulation No.: 45:I

LIQUID DISHWASHING DETERGENT

RAW MATERIALS	% By Weight
Water	51.9
BIO SOFT D-62	32.0
STEOL CS-460	10.0
NINOL 128-EXTRA	6.0
Preservative	0.1

Mixing Procedure:
   Adjust pH to 7.0-7.8
Properties:
   % active: 31.0

Formulation No.: 45:II

LIQUID DISHWASHING DETERGENT

RAW MATERIALS	% By Weight
Water	45.9
BIO SOFT D-62	33.0
STEOL CS-460	15.0
NINOL 128-EXTRA	6.0
Preservative	0.1

Mixing Procedure:
   Adjust pH to 7.0-7.8
Properties:
   % Active: 35.0

Formulation No.: 45-III

SOURCE: Stepan Co.: Suggested Formulations

LIQUID DISHWASHING DETERGENT (PREMIUM)

RAW MATERIALS   % By Weight

BIO SOFT LD-150   98.8
Sodium chloride   1.2

Mixing Procedure:
   Blend ingredients in order given. Adjust pH with sulfuric
acid.

Properties:
   Appearance   clear yellow liquid
   Viscosity @ 25C, cps   200
   pH, as is   6.3
   % solids   50.6
   Freeze/thaw (3 cycles)   Pass

Performance:
   Equal to "New" Joy by three test methods: Colgate Mini-Plate
test, Stepan DW-X test, and the "Pellet test."

SOURCE: Stepan Co.: Formulation No. 60

LIQUID DISHWASHING DETERGENT (MEDIUM)

RAW MATERIALS   % By Weight

Water   62.3
STEPANATE X   5.0
BIO SOFT LD-190   32.7

Mixing Procedure:
   Blend ingredients in order given.

Properties:
   Appearance   clear yellow liquid
   Viscosity @ 25C, cps   290
   pH, as is   8.4
   Solids, %   32.0
   Cloud point, C   <5
   Freeze/thaw (3 cycles)   Pass

Performance:
   Colgate Mini-Plates test: 22 mini-plates washed

SOURCE: Stepan Co.: Formulation No. 52

LIQUID DISHWASHING DETERGENT
Manual--High Sudsing
Anionic Type

RAW MATERIALS	% By Weight
Water	54.6
Sodium alkylbenzene sulfonate	20.4
Sodium xylene sulfonate	4.0
ALIPAL CO-436	18.0
GAFAMIDE CDD-518	3.0
	100.0

Perfume, colorants and opacifiers added, as desired, replacing water.

Manufacturing Procedure:
1. Add sodium alkylbenzene sulfonate to water and sodium xylene sulfonate. (Note: Alternatively, alkylbenzene sulfonic acid can be neutralized with sodium hydroxide to yield the same sodium alkylbenzene sulfonate activity.)
2. Add the remaining components in the order listed.

Physical Properties:
pH (as is)	8.7
pH (1%)	8.1
Viscosity	600 cps
Specific Gravity	1.02

SOURCE: GAF Corp.: Formulary: Prototype Formulation GAF 5201

HAND DISHWASHING DETERGENT

RAW MATERIALS	% By Weight
MIRANOL CS CONC.	15.0
Dodecylbenzene Sulfonic Acid	12.0
Sodium Hydroxide, 50%	3.0
Igepal CO-630	3.0
Cedemide CX	3.0
Water	64.0

SOURCE: Miranol Inc.: MIRANOL Products for Household/Industrial Applications: Suggested Formulation

DISHWASHING DETERGENT(24%)(2878-114)

RAW MATERIALS   % By Weight

Water	to 100
Caustic soda (50% sodium hydroxide)	3.3
Dodecylbenzene sulfonic acid (DDBSA)	13.0
TRYCOL 6953 POE (12) Nonylphenol	2.0
Sodium ethoxylated alcohol sulfate (60% active)	7.5
EMID 6514 Cocamide DEA	2.0
Sodium xylene sulfonate (SXS)	3.0
Citric acid (50%) (to pH 6.5-7.5)	q.s.
Fragrance, dye, opacifier and preservative	as desired

Blending Procedure:
   To the blending tank, add the water, caustic soda and DDBSA.
At this point, the pH should be greater than 5.0 to prevent
tank corrosion. If not, immediately add more caustic soda.
Add the remainder of the ingredients in the order listed. The
amount of EMID 6514 and SXS can be adjusted to obtain the desired
viscosity. Increase EMID 6514 to increase viscosity. Increase
SXS to decrease viscosity.
   Adjust the batch to final pH before the dye (as an aqueous
solution) is added.

SOURCE: Emery Chemicals: Specialty Chemicals Formulary:
         Formulation 2878-114

PEARLESCENT DISHWASHING DETERGENT(2878-115)

RAW MATERIALS   % By Weight

Water	to 100
Caustic soda (50% sodium hydroxide)	3.3
Dodecylbenzene sulfonic acid (DDBSA)	13.0
TRYCOL 5943 POE (12) Tridecyl Alcohol	2.0
EMID 6500 Cocamide MEA	1.0
EMEREST 2350 Glycol Stearate	1.0
EMERSAL 6453 Sodium Laureth Sulfate	15.0
Sodium xylene sulfonate (40%) (SXS)	3.0
Citric acid (50%) (to pH 6.5-7.5)	q.s.
Dye, preservative and fragrance	as desired

Blending Procedure:
   To the batch tank, add the water, caustic soda and DDBSA.
At this point, the pH should be greater than 5.0. If not, immediately add more caustic soda. Heat the batch to 150-170F. Add
the TRYCOL 5943, EMID 6500 and EMEREST 2350. Mix until the
EMEREST 2350 has completely dissolved. Cool to 110 F and add
the remaining ingredients. Adjust the batch to final pH before
the dye (as an aqueous solution) is added.
   Note: The viscosity of the finished product can be adjusted
by increasing or decreasing the SXS.

SOURCE: Emery Chemicals: Specialty Chemicals Formulary: 2878-115

DISHWASHING DETERGENT LIQUID, HAND
(Clear)

RAW MATERIALS	% Active
TRITON X-102 Surfactant or	
TRITON X-100 Surfactant	12.0
Sodium Linear Alkylate Sulfonate (60%)	23.0
Lauricdiethanolamide	3.0
Ethanol	2.0
Water	60.0
	100.0

Use Dilution: 1 oz./2 gal. water.

Lit. Ref: CS-407, CS-427

SOURCE: Rohm and Haas Co.: Specialty Chemicals: Detergent
Formulations for Industrial and Institutional Industry:
Lit. Ref.: CS-407, CS-427

DISHWASHING DETERGENT LIQUID, HAND
(Pink Opacified)

RAW MATERIALS	% Active
TRITON X-102	12.0
Sodium Linear Alkylate Sulfonate (60%)	23.0
Lauricdietnanolamide	3.0
Latex E-284 (40%) (Opacifier)	2.0
Calcozine Rhodamine BX Conc.	(0.0025)
Ethanol	2.0
Water	58.0
	100.0+

Use Dilution: 1 oz./2 gal. water

Lit. Ref: CS-407

SOURCE: Rohm & Haas Co.: Specialty Chemicals: Detergent
Formulations for Industrial and Institutional Industry:
Lit. Ref: CS-407

GENERIC DISHWASHING DETERGENT(14%)(2887-051)

RAW MATERIALS	% By Weight
Water	to 100
Caustic soda (50% sodium hydroxide)	2.5
Dodecylbenzene sulfonic acid (DDBSA)	10.0
EMERSAL 6453 Sodium Laureth Sulfate (28%)	9.0
EMID 6514 Cocamide DEA	1.0
Tetrasodium EDTA (40%)	0.5
Citric acid (50%) (to pH 6.5-7.5)	q.s.
Sodium chloride	1.0
Fragrance, dye, preservative and opacifier	as desired

Blending Procedure:

Charge water to batching tank, then add raw materials in the order listed. The pH of the batch tank should be higher than 5 after the DDBSA has been added. If not, adjust with the sodium hydroxide before proceeding.

A preblend of the sodium chloride with sufficient water to solubilize will facilitate the blending of this viscosity enhancer. One percent salt (NaCl) should give a viscosity greater than 100 cP at 25C (Brookfield Viscometer). Excessive use of salt should be avoided or the formula will not be stable.

Adjust the batch to final pH before the dye (as an aqueous solution) is added.

For an opaque, lotion-type product, add 0.15% of an opacifier. One part WITCOPAQUE R-25 (Witco) or E-288 (Morton) opacifier must be preblended with 3 parts water before addition to batch tank. Final pH must be adjusted before the opacifier solution is added.

SOURCE: Emery Chemicals: Specialty Chemicals Formulary: Formulation 2887-051

MEDIUM ACTIVE DISHWASHING DETERGENT(25%)(2887-051)

RAW MATERIALS	% By Weight
Water	64.7
Caustic soda (50% sodium hydroxide)	3.3
Dodecylbenzene sulfonic acid (DDBSA)	13.0
TRYCOL 6953 POE (12) Nonylphenol	2.0
EMERSAL 6453 Sodium Laureth Sulfate (28%)	15.0
EMID 6514 Cocamide DEA	2.0
Tetrasodium EDTA (40%)	0.5
Sodium xylene sulfonate (40%) (SXS)	q.s. (~3.0)
Citric acid (50%) (to pH 6.5-7.5)	q.s.
Dye, fragrance, preservative and opacifier	as desired
	100.0

Blending Procedure:

   Charge the water (warm, if possible) to the batching tank
then add the remaining raw materials in the order listed. The
pH of the mixture should be higher than 5 after the DDBSA has
been added. If not, adjust the pH with sodium hydroxide before
proceeding. Add sodium xylene sulfonate to adjust the formula
to approximately 150-300 cP (Brookfield Viscometer).

   Adjust the batch to the final pH before the dye (as an aqueous solution) is added. If an opaque, lotion-type product is
desired, 0.15% of an opacifier may be added. One part WITCOPAQUE
R-25 (Witco) or E-288 (Morton) opacifier must be preblended with
three parts water before its addition to the batch tank. The
final pH must be adjusted before the opacifier solution is
added.

SOURCE: Emery Chemicals: Specialty Chemicals Formulary:
        Formulation 2887-051

CONCENTRATED LIQUID MACHINE DISHWASHING COMPOUND

RAW MATERIALS	% By Weight
MIRANOL J2M CONC.	2.0
Potassium Hydroxide, 45%	9.0
Kasil #1	27.0
Sodium Gluconate	2.0
Tetrapotassium Pyrophosphate	14.0
Water	46.0

SOURCE: Miranol Inc.: MIRANOL Products for Household/Industrial
       Applications: Suggested Formulation

INSTITUTIONAL LIQUID MACHINE DISHWASHING COMPOUND

RAW MATERIALS	% By Weight
MIRANOL JEM CONC.	3.0
Tetrapotassium Pyrophosphate	10.0
Sodium Metasilicate Pentahydrate	10.0
Water	77.0

SOURCE: Miranol, Inc.: MIRANOL Products for Household/Industrial
       Applications: Suggested Formulation

MEDIUM DUTY LIQUID MACHINE SPRAY WASH

RAW MATERIALS	% By Weight
MIRANOL J2M CONC.	1.0
Potassium Hydroxide, 45%	20.0
Kasil #1	22.0
Gluconic Acid	8.0
Water	49.0

SOURCE: Miranol Inc.: MIRANOL Products for Household/Industrial
       Applications: Suggested Formulation

AUTOMATIC DISHWASHING DETERGENT (POWDER)

RAW MATERIALS	% By Weight
Sodium tripolyphosphate, dense	40.0
MAKON NF-5	3.0
Britesil H-20	30.0
CDB Clearon	2.0
Soda ash, dense	25.0

Mixing Procedure:
  Charge blender with soda ash. Add MAKON NF-5 and mix for one minute. Add remaining ingredients and mix for one minute.
Properties:
  Appearance   White free flowing powder
  pH, at 1%   11.0
Use Instructions:
  As per dishwasher manufacturer's instructions.
Performance:
  - excellent performance at low temperature
  - low foam & defoaming of food soils
  - no spotting, etching or streaking on glass wares
  - hardwater tolerance
Comments:
  MAKON NF-5 is an excellent low foaming and defoaming surfactant

SOURCE: STEPAN CO.: Formulation No. 22

MECHANICAL DISHWASHING DETERGENT
Chlorinated Type

RAW MATERIALS	% By Weight
ANTAROX BL-330	2.0
Sodium metasilicate 5-H2O	20.0
Clearon CDB	3.0
Sodium tripolyphosphate, powder	40.0
Sodium carbonate (lt. density)	35.0
	100.0

Manufacturing Procedure:
  1. Disperse ANTAROX BL-330 in sodium tripolyphosphate.
  2. Add sodium metasilicate 5-H2O, sodium carbonbate and
     Clearon CDB.
Physical Properties:
  pH (1%)   11.2
  Specific Gravity   .70

SOURCE: GAF Corp.: Formulary: Prototype Formulation GAF 5228

MECHANICAL DISHWASHING DETERGENT
Liquid Type

RAW MATERIALS   % By Weight

IGEPAL CO-730 (1% active)	0.5
GANTREZ AN-149	1.0
Potassium hydroxide (50% active)	3.0
Sodium silicate (46% active)	17.0
ANTAROX BL-330	3.0
Tetrapotassium pyrophosphate (60% active)	63.3
Water	12.2
	100.0

Manufacturing Procedure:
1. Dissolve GANTREZ AN-149 in IGEPAL CO-730/water mixture by stirring at room temperature for eight hours. Heat to 45-50C if not totally in solution.
2. Maintain heat at 45C and add rest of ingredients individually.

Physical Properties:
pH (as is)	13.2
pH (1%)	10.4
Viscosity	1060 cps
Specific Gravity	1.14

SOURCE: GAF CORP.: Formulary: Prototype Formulation GAF 5233

MECHANICAL DISHWASHING DETERGENT
Chlorinated Type

RAW MATERIALS   % By Weight

ANTAROX BL-330	2.0
Sodium tripolyphosphate, powder	48.0
Sodium metasilicate 5-H2O	20.0
Trisodium phosphate, chlorinated	30.0
	100.0

Manufacturing Procedure:
1. Disperse ANTAROX BL-330 onto sodium tripolyphosphate. Mix well to avoid lumping of surfactant and powder.
2. Add sodium metasilicate 5-H20. Add trisodium phosphate, chlorinated. Mix thoroughly.

Physical Properties:
pH (1%)	11.4
Specific Gravity	.78

SOURCE: GAF CORP.: Formulary: Prototype Formulation GAF 5227

DISHWASHING DETERGENT, MACHINE
(Chlorine-Releasing Agent)

RAW MATERIALS	% By Weight
TRITON CF-54 Surfactant	2.0
Sodium Carbonate	41.0
Sodium Tripolyphosphate (STPP)	30.0
Sodium Metasilicate (Anhydrous)	25.0
Sodium Dichloro-s-triazinetrione (CDB Clearon)	2.0
	100.0

Mixing Instructions:
   Thoroughly blend TRITON CF-54 Surfactant and soda ash, STPP, and sodium metasilicate. Add chlorine-release agent.

Direction for Use:
   Add 2-4 tablespoons per load in household dishwashers.

SOURCE: Rohm and Haas CO.: Specialty Chemicals: Detergent
        Formulations for Industrial and Institutional Industry:
        Lit. Ref.: CS-412

DISHWASHING DETERGENT, MACHINE

RAW MATERIALS	% By Weight
Water	0.89
ACRYSOL ASE-108 Stabilizer	11.10
Potassium Hydroxide (50% solution)	40.00
Tetrapotassium Pyrophosphate (anhydrous)	20.00
Potassium Silicate (40% solution)	25.00
TRITON CF-32 Surfactant (95% active)	3.00
Dye	0.01

Mixing Instructions:
   Add the components slowly in the order listed with subsurface agitation. Be sure each ingredient is completely solubilized or dispersed before adding the next one. A mixture that provides turbulence is recommended, but vortices should be avoided. Variations on the formulation are possible, but compatibility and product stability should be carefully evaluated.

Percent Solids	54.9
pH (50% aqueous solution)	11.8
Bulk Density, lbs./gal.	12.23
Specific Gravity @ 25C	1.47
Viscosity, cps @ 25C	960

SOURCE: Rohm & Haas Co.: Specialty Chemicals: Detergent
        Formulations for Industrial and Institutional Industry:
        Lit. Ref: CS-432, CS-500

DISHWASHING DETERGENT, MACHINE

RAW MATERIALS   % By Weight

Water   43.53
ACRYSOL ASE-108 Stabilizer   6.9
Potassium Hydroxide (45% solution)   1.56
Tetrapotassium Pyrophosphate (anhydrous)   40.00
Tetrasodium Ethylenediaminetetraacetate   5.00
TRITON CF-32 Surfactant (95% active)   3.00
Dye   0.01

Percent Solids   49.8
pH (0.5% aqueous solution)   10.4
Bulk Density, lbs./gal.   11.95
Specific Gravity @ 25C   1.43
Viscosity cps. @ 25C   280

SOURCE: Rohm and Haas Co.: Specialty Chemicals: Detergent
        Formulations for Industrial and Institutional Industry:
        Lit. Ref: CS-432/CS-500

DISHWASHING DETERGENT, MACHINE-A
(Chlorine-Releasing Agent)

RAW MATERIALS   % By Weight

TRITON CF-10 Surfactant or
   TRITON CF-32 Surfactant   3.0
Sodium Tripolyphosphate (STPP)   50.0
Sodium Metasilicate (Anhydrous)   45.0
Sodium Dichloro-s-triazinetrione (CDB Clearon)   2.0
                                                              100.0

Mixing Instructions:
    Thoroughly blend TRITON CF-detergent with builders.
    Add chlorine-release agent.

Directions for Use:
    Add 2-4 tablespoons per load in household dishwashers.

SOURCE: Rohm and Haas Co.: Specialty Chemicals: Detergent
        Formulations for Industrial and Institutional Industry:
        Lit. Ref: CS-432/CS-436

DISHWASHING DETERGENT, MACHINE-B
(Chlorine-Releasing Agent)

RAW MATERIALS	% By Weight
TRITON CF-10 Surfactant or	
TRITON CF-32 Surfactant	2.0
Sodium Tripolyphosphate (STPP)	34.0
Sodium Metasilicate (Anhydrous)	40.0
Soda Ash	22.0
Sodium Dichloro-s-triazinetrione (CDB Clearon)	2.0
	100.0

Mixing Instructions:
   Thoroughly blend TRITON CF-detergent with builders.
   Add chlorine-release agent.
Directions for Use:
   Add 2-4 tablespoons per load in household dishwashers.
   Less costly version.

SOURCE: Rohm & Haas Co.: Specialty Chemicals: Detergent
         Formulations for Industrial and Institutional Industry:
         Lit. Ref: CS-432, CS-436

DISHWASHING DETERGENT, MACHINE(PHOSPHATES/SILICATE/SURFACTANT)

RAW MATERIALS	% By Weight
Water	50.49
ACRYSOL ASE-108 Stabilizer	6.90
Potassium Hydroxide (45% solution)	1.27
Tetrapotassium Pyrophosphate (anhydrous)	20.00
Trisodium Phosphate (anhydrous)	5.00
Sodium Silicate (37.5%)	13.33
TRITON CF-32 Surfactant (95% active)	3.00
Dye	0.01

Properties:

Solids, %	34.68
pH (0.5% Aq. Soln.)	11.0
Bulk Density, lb./gal.	10.85
Specific Gravity @ 25C	1.30
Viscosity cps @ 25C	380

Mixing Procedure:
   Slowly add ingredients in listed order with subsurface agitation in a vessel equipped with baffles and a 4-blade turbine
impeller. Avoid high-speed agitation and vortex formation which
may entrap air bubbles. Dissolve or disperse each ingredient
completely before adding.

SOURCE: Rohm and Haas Co.: Specialty Chemicals: Detergent
         Formulations for Industrial and Institutional Industry:
         Lit. Ref.: CS-432, CS-500

DISHWASHING DETERGENT, MACHINE/BIODEGRADABLE-B
(Chlorine-Releasing Agent)

RAW MATERIALS	% By Weight
TRITON DF-18 Surfactant (90%)	2.2
Soda Ash	41.0
Sodium Metasilicate (Anhydrous)	24.8
Sodium Tripolyphosphate (STPP)	30.0
Sodium Dichloro-s-trazinetrione (CDB Clearon)	2.0
	100.0

Mixing Instructions:
    Thoroughly blend TRITON DF-18 Surfactant with builders.
    Add chlorine-release agent.

Direction For Use:
    Add 2-4 tablespoon per load in household dishwashers.

SOURCE: Rohm and Haas Co.: Specialty Chemicals: Detergent
        Formulations for Industrial and Institutional Industry:
        Lit. Ref.: CS-405

MACHINE DISHWASHING DETERGENT

RAW MATERIALS	% By Weight
Sodium tripolyphosphate	35
Sodium carbonate	10
PLURAFAC RA-43 surfactant or	
PLURONIC 25R2 polyol	3
Sodium metasilicate	25
Sodium sulfate	25
Chlorinated isocyanurate	2

SOURCE: BASF Corp.: Cleaning Formulary: Formulation #3500

DISHWASHING DETERGENT, MACHINE

RAW MATERIALS   % By Weight

Water   55.76
ACRYSOL ASE-108 Stabilizer   6.90
Potassium Hydroxide (45% solution)   1.33
Tetrapotassium Pyrophosphate (anhydrous)   25.00
Trisodium Phosphate (anhydrous)   5.00
Sodium Nitrilotriacetate*   3.00
TRITON CF-32 Surfactant (95% Active)   3.00
Dye   0.01

   * Sequestrant for use with hard water

Properties:
   Solids Content, %   37.7
   pH (0.5% aq. soln)   10.8
   Specific Gravity @ 25C   1.33
   Bulk Density, lb./gal   11.1
   Viscosity cps/25C   180
   Use Level   2-4 tbsp./gal.
Mixing Procedure:
   Slowly add ingredients in listed order with subsurface agit-
ation in a vessel equipped with baffles and a 4-blade turbine
impeller. Avoid high-speed agitation and vortex formation which
may entrap air bubbles. Dissolve or disperse each ingredient
completely before adding the next one. Disperse dye in a small
amount of water withheld from the initial charge.

SOURCE: Rohm and Haas Co.: Specialty Chemicals: Detergent
        Formulations for Industrial and Institutional Industry:
        Lit. Ref: CS-432, CS-500

DISHWASHING DETERGENT, MACHINE/BIODEGRADABLE

RAW MATERIALS   % By Weight

TRITON DF-18 Surfactant (90%)   2.2
Sodium Hydroxide   41.0
Sodium Metasilicate (Anhydrous)   24.8
Sodium Tripolyphosphate (STPP)   30.0
Sodium Dichloro-s-triazinetrione (CDB Clearon)   2.0
                                                                   100.0

Mixing Instructions: Thoroughly blend TRITON DF-18 Surfactant
                   with builders. Add chlorine-release agent.
Directions for use:
   Add 2-4 tablespoon per load in household dishwashers.

SOURCE: Rohm and Haas Co.: Specialty Chemicals: Detergent
        Formulations for Industrial and Institutional Industry:
        Lit. Ref.: CS-405

INDUSTRIAL DISHWASH FORMULATION: REGULAR DUTY, GOOD QUALITY, MILD LIQUID

RAW MATERIALS	% By Weight
NEODOL 91-6	6.8
NEODOL 91-2.5	2.3
DDBSA (98%)	5.0
FADEA	0.8
Sodium hydroxide (50%)	1.3
Sodium xylene sulfonate (40%)	8.0
Sodium tripolyphosphate, anhydrous basis	1.8
Sodium hexametaphosphate	0.6
Water, dye, perfume, preservatives	to 100%

Properties:

Viscosity, 73F, cps	45
Clear point, F	37
pH	7.4

INDUSTRIAL DISHWASH FORMULATION: GOOD QUALITY POT AND PAN CLEANER

RAW MATERIALS	% By Weight
NEODOL 91-6	10.5
NEODOL 91-2.5	4.5
DDBSA (98%)	8.8
FADEA	3.3
Sodium hydroxide (50%)	2.2
Sodium hexametaphosphate	2.0
Sodium xylene sulfonate (40%)	5.0
Water, dye, perfume, preservatives	to 100%

Properties:

Viscosity, 73F, cps	83
Clear point, F	37
pH	8.6

INDUSTRIAL DISHWASH FORMULATION: ALKALINE POT AND PAN CLEANER

RAW MATERIALS	% By Weight
NEODOL 91-6	8.6
NEODOL 91-2.5	2.9
NEODOL 25-3S (60%)	2.9
DDBSA (98%)	5.9
FADEA	3.5
Diethanolamine	2.6
Sodium hexametaphosphate	2.0
Water, dye, perfume, preservatives	to 100%

Properties:

Viscosity, 73F, cps	122
Clear Point, F	34
pH	9.1

SOURCE: Shell Chemical Co.: NEODOL Formulary: Formulations

LIQUID MACHINE DISHWASHING CONCENTRATE

RAW MATERIALS	% By Weight
MIRANOL J2M CONC.	1.5
Potassium Hydroxide, 45%	7.0
Kasil #1	20.0
Tetrapotassium Pyrophosphate	10.5
Water	61.0

SOURCE: Miranol Chemical Co.: MIRANOL Products for Household/
    Industrial Applications: Suggested Formulation

MACHINE DISHWASHING DETERGENT

RAW MATERIALS	% By Weight
ALKAWET CF	2.0
Sodium carbonate	18.0
Sodium metasilicate, anhydrous	30.0
Sodium tripolyphosphate	50.0

SOURCE: Lonza: Product Information: ALKAWET CF: Formulation
    C-99-91

AUTOMATIC DISHWASHER DETERGENT
Powder

Raw Materials	% By Weight
SURFONIC LF-17	2
Sodium Metasilicate, anhydrous	37
Sodium Tripolyphosphate	37
Sodium Carbonate	25

SOURCE: Texaco Chemical Co.: Suggested Formulation

MACHINE DISHWASHING FORMULATION

RAW MATERIALS	% By Weight
MAZER MACOL 40	3.0
Tetrasodium Pyrophosphate	35.0
Sodium Tripolyphosphate	20.0
Sodium Metasilicate, Pentahydrate	10.0
Chlorinated Cyanurate	2.0
Sodium Carbonate	18.0
Water*	12.0

Procedure:
1. Spray a mixture of the MACOL 40 and water* onto the Tetra-sodium pyrophosphate (alone or mixed with the other anhydrous inorganic salts) while continually mixing, whereby hydration and simultaneous absorption of the surfactant occur.
2. Add the hydrated Sodium Metasilicate to the hydrated phosphates, still mixing.
3. Add the Chlorinated Cyanurate to the mixture and continue mixing until homogenized and a dry, free-flowing, granular product is obtained. Chlorine-containing compounds which may also be used in this process include trisodium phosphate (chlorinated), trichlorocyanuric acid, salts of di-chlorocyanuric acid, dichlorodimethylhydantoin (Halane) and other organic chlorine-active compounds.

Maximum chlorine stability is obtained when the MACOL 40, diluted with water, is completely absorbed by the phosphate mixture. Step 1 avoids temperature control problems and eliminates the need for an aging period.

In Step 2, hydrated Metasilicate does not promote agglomeration. It is essential that the Metasilicate be added <u>after</u> the phosphate mixture has hydrated and absorbed the surfactant, to avoid discoloration and degradation.

* The amount of water used should not exceed that shown in the formulation.

SOURCE: Mazer Chemicals, Inc.: Household/Industrial T-20B: Formulation 1

MACHINE DISHWASHING POWDER

RAW MATERIALS	% By Weight
MIRANOL JEM CONC. or	
MIRANOL J2M CONC.	0.7
Nitrilotriacetic Acid	41.0
Sodium Metasilicate Pentahydrate	55.3
Sodium Chloroisocyanurate	2.0
Polyethylene Glycol	1.0

SOURCE: Miranol Chemical Co.: MIRANOL Products for Household/ Industrial Applications: Suggested Formulation

3. Disinfectants

DISINFECTANT CLEANER

RAW MATERIALS	% By Weight
REWOTERIC AM-V	20.0
1,2 Propyleneglycol	6.0
Ingasan DP 300	0.5
REWOQUAT B 50	5.0
Tetrapotassium Pyrophosphate	6.0
Trilon B, liquid	1.0
Water	qs100

Mixing Procedure:
    Mix ingredients with water in order shown.

SOURCE: Sherex: Industrial Formulation 5:05.4

GENERAL DISINFECTANT SOLUTION

RAW MATERIALS	% By Weight
VARION CDG	17.0
VARIQUAT 50MC	10.0
Glyoxal (40%)	5.0
Water	68.0

Mixing Procedure:
    Dissolve the 50MC into the water and add the Glyoxal and the CDG.

SOURCE: Sherex: Industrial Formulation 6:05.4

GENERAL DISINFECTANT

RAW MATERIALS	% By Weight
VARION CDG	17.0
VARIQUAT 50MC	10.0
Glyoxal (40%)	5.0
Water	qs100

Mixing Procedure:
    Add 50MC and glycol to water followed by the CDG.

SOURCE: Sherex: Industrial Formulation 10:01.7

DISINFECTANT CLEANERS*
LIQUID CONCENTRATE WITH PHOSPHATE

RAW MATERIALS   % By Weight

NEODOL 25-9   7.5
Bardac 22 (50%)   10.0
Tetrapotassium pyrophosphate   8.0
Water, dye, perfume   to 100%

Properties:
   Viscosity, 73F, cps   36
   Phase coalescence temp., F   158
   pH   10.6
Use Concentration: 1-2 oz/gal.

HOSPITAL/CLINICAL STRENGTH
RAW MATERIALS   % By Weight

NEODOL 25-9   7.0
Bardac 22 (50%)   15.0
EDTA   6.0
Sodium carbonate   3.0
Barquat MD-50 (50%)   10.0
Sodium xylene sulfonate (40%)   5.0
Water   to 100%

Properties:
   Viscosity, 73F, cps   31
   Phase coalescence temp., F   >176
   pH   9.7
Use Concentration:
   For hospital use: 2 oz/gal.
   For institutional, non-hospital use: 1 oz/gal.

NON-PHOSPHATE LIQUID CONCENTRATE
RAW MATERIALS   % By Weight

NEODOL 25-9   6.0
Bardac 22 (50%)   9.0
EDTA   5.0
Sodium carbonate   1.0
Sodium metasilicate, pentahydrate   0.9
Sodium xylene sulfonate (40%)   5.0
Pine oil   5.0
Water, dye, perfume   to 100%
Properties:
   Viscosity, 73F, cps   15
   Phase coalescence temp., F   >176
   pH   12.0
Use Concentration: 2 oz/gal.
  * Cleaners making a disinfectant claim require EPA
    registration.

SOURCE: Shell Chemical Co.: The NEODOL Formulary: Formulations

HOSPITAL DISINFECTANT

RAW MATERIALS	% By Weight
Water	84.2
VARIQUAT 5O MC	9.6
VAROX 365	4.0
Versene 100	1.5
Sodium Hydroxide	0.7

SOURCE: Sherex: Industrial Formulation 35:5.4.2

SANITIZER--LIQUID, ACID

RAW MATERIALS	% By Weight
HOSTAPUR SAS 60	7,5
GENAMINOX CS (amine oxide; 30%)	5,0
Nonylphenol + 9 EO	5,0
Hydrochloric acid (37%)	10,0
Phosphoric acid (85%)	10,0
Water, parfume, dyestuff	ad 100,0

Production procedure:
    HOSTAPUR SAS 60, GENAMINOX CS and Nonylphenol + 9 EO will be dissolved in water. Afterwards one should add phosphoric acid, hydrochloric acid, parfume and dyestuff.
Tests
    Appearance at 20C   cloudy, viscous, liquid
    Viscosity   400 mPas
    pH-Value (1%)   1,0
    Cloud point (-5C)   cloudy, stable
    Stability (4 Mon./40C)
    Freeze and thaw-test   cloudy, stable

SOURCE: Hoechst/Celanese: Formulation D-6016

SANITIZER(ALSO FOR SPRAY)--LIQUID, CLEAR

RAW MATERIALS	% By Weight
HOSTAPUR SAS 30	10.0
GENAMINOX KC (amine oxide, 30%)	10.0
Sodium hydroxide (100 %ig)	1.0
Medialan LD	2.5
Sodium hypochlorite (150 g/l active)	50.0
LOSER GX 5 (Alkylarylpolyglycolether)	5.0
Water	21.5

Tests:
    pH-value (1%)   11.3
    Viscosity   48 mPas
    stability (-5C)   clear

SOURCE: Hoechst/Celanese: Formulation D-6018

SANITIZER: LIQUID, CLOUDY, CONTAINS CHLORINE

RAW MATERIALS   % By Weight

HOSTAPUR SAS 60   2,5
GENAMINOX CS (amine oxide; 30%)   3,3
Sodiumhydroxid (100%)   1,0
Sodiumhypochlorite (150 g/l active)   50,0
Water, parfume   43,2

Production procedure:
   HOSTAPUR SAS, GENAMINOX CS, parfume and NaOCl will be
dissolved in cold water and the other ingredients will be
added afterwards.

Note:
   This solution is cloudy but stable!
   Offers an excellent chlorine stability.

Tests:
   pH-value (10%)   12,5
   Viscosity   40 mPas
   Freeze and thaw test   o.k.
   Cloud point (-5C)   cloudy, stable
   Stability (4 Mon/40C)   o.k.

SOURCE: Hoechst/Celanese: Formulation D-6009

SANITIZING CLEANER(LIQUID, 12.5% A.M.)

RAW MATERIALS   % By Weight

HOSTAPUR SAS 60   15.0
Nonylphenolethoxilate (6 EO)   3.0
Lauryldimethylamine Oxide (30%)   5.0
Phosphoric Acid (85%)   10.0
Water   AD100.0

Production Procedure:
   Dissolve HOSTAPUR SAS and nonionic in water. Add phosphoric
acid and GENAMINOX CS.

Tests:
   pH Value (10% Aqueous Solution T.Q.)   1.7
   Viscosity   800 MPAS

SOURCE: Hoechst/Celanese: Formulation D-6008

4. Floor Cleaners and Wax Strippers

FLOOR CLEANER

RAW MATERIAL   % By Weight

DOWANOL PM glycol ether   14
water   58
oleic acid   16
triethanolamine   12

Mix in order listed.

SOURCE: Dow Chemical U.S.A.: The Glycol Ethers Handbook:
        Suggested Formulation

FLOOR CLEANER

RAW MATERIALS   % By Weight

Water   58.0
Triethanolamine   12.0
M-PYROL   14.0
Oleic Acid   16.0
                                                                100.0

Manufacturing Procedure:
    Add oleic acid to water and, slowly, the rest of the compon-
ents.

Physical Properties:
     pH (as is)   8.2
     pH (1%)   8.0
     Viscosity   30 cps
     Specific Gravity   1.01

SOURCE: GAF Corp.: Formulary: Prototype Formulation GAF 5352

FLOOR CLEANER

RAW MATERIALS   % By Weight

Sodium carbonate   2.0
TRYCOL 6964 POE (9) Nonylphenol   6.0
Acrysol ICS-1   4.0
Sodium xylene sulfonate (SXS)   5.0
Ammonia (28 Be)   1.2
Fragrance and dye   as desired
Water (soft)   to 100

Blending Procedure:
   Add the water to the blending tank. While mixing, add the
ingredients to the blending tank in the order listed. Mix
until uniform.

SOURCE: Emery Chemicals: Specialty Chemicals Formulary:
        Formulation 2886-120

FLOOR CLEANER
(Stable to 120F)

RAW MATERIALS   % By Weight

MAZER MACOL NP 9.5   7
Sodium Tripolyphosphate   8
Tri Sodium Phosphate   2
Na Metasilicate Pentahydrate   2
MAZER MAPHOS 60A   2
Water   79

SOURCE: Mazer Chemicals, Inc.: Household/Industrial T-20B:
        Formulation 7

FLOOR CLEANER
cloudy, liquid

RAW MATERIALS	% By Weight
HOSTAPUR SAS 60	10,0
Nonylphenol + 4 EO	5,0
Oleic acid	2,0
Triethanolamin	2,0
White spirit	30,0
Terpentinoil	10,0
EDTA	2,0
Water, dyestuff, perfume	39,0

Production process:
HOSTAPUR SAS 60, NP-4 and EDTA will be mixed with perfume before adding oleic acid, TEA, white spirit, terpentinoil and water. This solution is cloudy but stable.

Tests:

pH-value (10 %ig)	8,9
Viscosity	200 mPas
Freeze and thaw test	o.k.
Stability (4 Mon./40C)	cloudy but stable

SOURCE: Hoechst/Celanese: Formulation D-4006

FLOOR CLEANER

RAW MATERIALS	% By Weight
Water	82.28
Acrysol ASE-108 Polymer (18%)	4.72
Sodium Hydroxide (10%)	2.80
Sodium Nitrite	0.15
Sodium Pentachlorophenate (Mitrol G-ST)	0.05
TRITON X-100 Surfactant	10.00
	100.00

Mixing Procedure:
Mix ACRYSOL ASE-108 Thickener with 1 part water and add to the rest of the water. Add caustic slowly with agitation, then other ingredients in listed order and mix completely.
Note: Cleaners containing a mixture of sodium salts and nonionic surfactant may discolor white vinyl tiles. Avoid excessive cleaner concentrations and prolonged contact.

Properties:
Solids Content, %   11.33
Appearance: Opaque off-white viscous liquid
Use Level: 1 oz./4 gallons water

SOURCE: Rohm and Haas Co.: Specialty Chemicals: Detergent Formulations for Industrial and Institutional Industry: Lit. Ref.: CS-427, CS-500

FLOOR CLEANER
Concentrate

RAW MATERIALS   % By Weight

GAFAC LO-529   6.0
Sodium tripolyphosphate   3.0
Tetrasodium pyrophosphate   3.0
Water   88.0
                                                            100.0

Manufacturing Procedure:
  1. Dissolve solids in the total amount of water.
  2. Add GAFAC LO-529. Mix thoroughly.

Physical Properties:
  pH (as is)   13.9
  pH (1%)   8.9
  Viscosity   60 cps
  Specific Gravity   1.02

SOURCE: GAF Corp.: Formulary: Prototype Formulation GAF 5354

FLOOR CLEANER

RAW MATERIAL   % By Weight

GAFAC LO-529   5.0
GAFAMIDE CDD-518   1.0
Tetrasodium pyrophosphate (60% active)   2.0
Trisodium phosphate   1.0
Diethylene glycol monoethyl ether   1.0
Water   90.0
                                                            100.0

   Perfume and colorants added, as desired, replacing water.

Manufacturing Procedure:
  1. Dissolve tetrasodium pyrophosphate in water. Add tri-
     sodium phosphate. Dissolve thoroughly.
  2. Add GAFAC LO-529 and GAFAMIDE CDD-518.
  3. Add solvent.

Physical Properties:
  pH (as is)   11.0
  pH (1%)   9.0
  Viscosity   290 cps
  Specific Gravity   1.01

SOURCE: GAF Corp.: Formulary: Prototype Formulation GAF 5353

FLOOR CLEANER
Liquid, Opaque

RAW MATERIALS	% By Weight
Wax (HOECHST-WACHS VP KST)	20.0
HOSTAPUR SAS 60	3.0
Nonylphenolethoxilate (10 EO)	1.0
Isotridecylalcohol Ethoxilate (8 EO)	2.0
Polyoxipropylen-Polyoxiethylen Condensate (GENAPOL PF 20)	2.0
EDTA-Solution	1.0
Methyl Glycole	1.0
Water (Hot, 60C)	39.0
Water (Cold), Dye, Perfume Oil	31.0

Production Procedure:
    Melt HOECHST-WACHS VP KST and dissolve it in hot water.
    Add HOSTAPUR SAS, Nonylphenolethoxilate, Isotridecylalcohol-
    Ethoxilate, Polyoxipropylen-Polyoxiethylen Condensate and the
    EDTA solution. Mix Methyl Glycole with the cold water and add
    it to the cooled off solution. Dye and perfume will be added
    in the end.
Tests:
    pH Value (10% Aqueous Solution T.Q.)   10.4
    Viscosity   Approx. 65 MPAS
    Stability (-5C)   Opaque
    Freeze and Thaw Test   O.K.

SOURCE: Hoecnst/Celanese: Formulation D-4004

FLOOR CLEANER (GARAGE)

RAW MATERIALS	% By Weight
TRITON X-114 Surfactant	2.5
Tallow Soap (Excelsior Soap)	2.0
Dipropylene Glycol Methyl Ether (Dowanol DPM)	6.5
Trisodium Phosphate (TSP)	3.0
Sodium Metasilicate Pentahydrate	3.0
Water	83.0
	100.0

Mixing Instructions:
    Stir TRITON X-114 and Dowanol DPM until homogeneous.
    Add tallow soap, then water after soap dissolves.
    Add TSP and metasilicate.
Directions for Use:
    Apply directly to oil spots. Allow 5 minutes to penetrate.
    Flush with water.

SOURCE: Rohm and Haas Co.: Specialty Chemicals: Detergent
        Formulations for Industrial and Institutional Industry:
        Lit. Ref.: CS-409

FLOOR CLEANER-LOW FOAM MACHINE APPLIED*

RAW MATERIALS   % By Weight

NEODOL 91-6   3.0
Triton H-66   4.0
Tetrapotassium pyrophosphate   8.0
Potassium hydroxide (45%)   1.5
Water, dye, perfume   to 100%

Properties:
   Viscosity, 73F, cps   5
   Phase coalescence temp., F   137
   pH   13.4

Use Concentration: 1-2 oz/4 gal.

  * CAUTION: This solution may discolor vinyl tiles if allowed
       to remain for lengthy periods.

SOURCE: Shell Chemical Co.: NEODOL Formulary: Suggested
      Formulation

FLOOR CLEANER, LOW-FOAM MACHINE SCRUBBER

RAW MATERIALS   % By Weight

TRITON DF-12 Surfactant   4.0
Tetrapotassium Pyrophosphate (TKPP)   8.0
TRITON H-66 Surfactant (50%)   10.0
Water   78.0
                                                      100.0

Appearance: Clear solution to 40C (104F.)

SOURCE: Rohm and Haas Co.: Specialty Chemicals: Detergent
      Formulations for Industrial and Institutional Industry:
      Lit. Ref.: CS-415, CS-433

FLOOR CLEANER, LOW-FOAM MACHINE SCRUBBER

RAW MATERIALS	% By Weight
TRITON X-114 Surfactant	3.0
TRITON H-66 Surfactant (50%)	4.0
Tetrapotassium Pyrophosphate (TKPP)	8.0
Potassium Hydroxide (45%)	2.5
Water	82.5
	100.0

Use Dilution: 1 to 2 oz./4 gal. water

SOURCE: Rohm & Haas Co.: Specialty Chemicals: Detergent
        Formulations for Industrial and Institutional Industry:
        Lit. Ref.: CS-409, CS-433

FLOOR CLEANER, LOW-FOAM MACHINE SCRUBBER

RAW MATERIALS	% By Weight
Water	53.27
ACRYSOL ASE-108 Thickener (18%)	8.33
Potassium Hydroxide (45%)	3.4
Tetrapotassium Pyrophosphate (Anhydrous)	22.0
Trisodium Phosphate (Anhydrous)	5.0
Potassium Silicate (Kasil #6)	5.0
TRITON DF-12 Surfactant	3.0
	100.0

Mixing Procedure: Add in listed order.

Properties:
    Solids Content, %   38.83
    Appearance   Milky emulsion
    Viscosity cps 25C   500 (3 days after preparation)
    Use Level   1-2 oz./4 gallons water

SOURCE: Rohm and Haas Co.: Specialty Chemicals: Detergent
        Formulations for Industrial and Institutional Industry:
        Lit. Ref.: CS-415, CS-500

FLOOR CLEANER WITH FBR

RAW MATERIALS   % By Weight

1)Sodium tripolyphosphate (STP)   2.2
Tetrapotassium Pyrophosphate   2.2
Water   95.6

2) REWOPOL FBR   7.0

Mixing Procedure
   Stir in STP and Pyrophosphate into water. Stir solution until
clear. Stir in FBR using good agitation.

SOURCE: Sherex Chemical Co.: Industrial Formulation 28:1.3

FLOOR CLEANER-GARAGE FLOOR

RAW MATERIALS   % By Weight

NEODOL 91-6   5.0
Triton H-66   4.0
Butyl OXITOL   6.5
Trisodium phosphate, anhydrous basis   1.3
Sodium metasilicate, pentahydrate   3.0
Water, dye, perfume   to 100%

Properties:
   Viscosity, 73F, cps   6
   Phase coalescence temp., F   126
   pH   13

Blending Procedure for Garage Floor Only:
   Add builders last with vigorous mixing until homogeneous.

SOURCE: Shell Chemical Co: NEODOL Formulary: Suggested
           Formulation

FLOOR CLEANER/WAX STRIPPER

RAW MATERIALS   % By Weight

Water   83
KLEARFAC AA-270 surfactant   6
Tetrapotassium pyrophosphate   9
Potassium hydroxide (45%)   2

Suggested use concentration: 2-4 oz. per gallon of water

SOURCE: BASF Corp.: Cleaning Formulary: Formulation #3476

FLOOR CLEANER/WAX STRIPPER*

RAW MATERIALS   % By Weight

Water   74.0
Sodium Hydroxide (50%)   3.4
Triton QS-44 Surfactant (80%)   3.2
Sodium Metasilicate, Anhydrous   0.4
Tetrapotassium Pyrophosphate (TKPP)   12.0
Tetrasodium Pyrophosphate (TSPP)   7.0
                                                          100.0

   * Addition of 1% Ammonium Hydroxide will allow easy removal
     of polymer finish from most types of floors.

Mixing Instructions:
    Mix in listed order, agitating to complete uniformity before
adding next ingredient.

Properties:
    Appearance   clear
    pH   13.4
    Viscosity, cps.   10

Use Dilution:
    Cleaning - 4 oz./gal. water
    Stripping - 16 oz./gal. water

SOURCE: Rohm and Haas Co.: Specialty Chemicals: Detergent
        Formulations for Industrial and Institutional Industry:
        Lit. Ref.: CS-410

FLOOR CLEANER AND WAX STRIPPER

RAW MATERIALS   % By Weight

MAZER MACOL 25   10
Oleic Acid   5
MEA   5
Tetrasodium Pyrophosphate   5
Water   75

SOURCE: Mazer Chemicals, Inc.: Household/Industrial T-20B: 6

FLOOR CLEANERS AND WAX STRIPPER

RAW MATERIAL   % By Weight

SURFONIC N-95   1
Monoethanolamine   1-2
Alkaline builders   As desired
Water, dye, fragrance   As desired*

FLOOR CLEANER AND WAX STRIPPER

RAW MATERIALS   % By Weight

SURFONIC N-85   8
Tetrapotassium** pyrophosphate   2
Trisodium phosphate   1
Water, dye   As desired*

* These products should be diluted with water 20 to 1 when used
  as a wax stripper and 80 to 1 when used for general cleaning.
** Or tetrasodium

The detergency and wetting properties of SURFONIC N-95 are
utilized in industrial floor cleaners and wax strippers.
Inclusion of monoethanolamine provides wax stripping capability.

SOURCE: Texaco Chemical Co.: SURFONIC N-Series Surface-Active
        Agents: Suggested Formulations

FLOOR CLEANER/WAX STRIPPER*

RAW MATERIALS   % By Weight

Water	77.0
Dipropylene Glycol Methyl Ether (DOWANOL DPM)	4.0
Monoethanolamine	0.5
TRITON N-101 Surfactant or	
TRITON X-100 Surfactant	5.0
Tetrasodium Ethylenediaminetetraacetate (VERSENE 100-39%)	6.0
Sodium Hydroxide (50%)	1.0
TRITON QS-30 Surfactant (90%)	3.0
Sodium Metasilicate, Anhydrous	3.5
	100.0

* Addition of 1% Ammonium Hydroxide will allow easy removal of polymer finish from most types of floors.

Mixing Instructions:
   Mix in listed order agitating to complete uniformity before adding next ingredient.

Properties:
Appearance	clear
pH	13.3
Viscosity, cps.	10

Use Dilution:
   Cleaning: 8 oz./gal. water
   Stripping:16 oz./gal. water

SOURCE: Rohm and Haas Co.: Specialty Chemicals: Detergent
        Formulations for Industrial and Institutional Industry:
        Lit. Ref.: CS-408, CS-427, CS-439

FLOOR WAX STRIPPER

RAW MATERIALS   % By Weight

DOWANOL PM glycol ether	8.0
trisodium phosphate	2.5
Triton X-100 surfactant	8.5
water	81.0

   Dissolve trisodium phosphate in water. Then mix in the Triton X-100 followed by DOWANOL PM.

SOURCE: Dow Chemical U.S.A.: The Glycol Ethers Handbook:
        Suggested Formulation

FLOOR CLEANER/WAX STRIPPER-POWDER
GOOD QUALITY LIQUID

RAW MATERIALS   % By Weight

NEODOL 91-6	10.0
Tetrapotassium pyrophosphate	5.0
Monoethanolamine	5.0
Oleic acid	5.0
Water, dye, perfume	to 100%

Properties:

Viscosity, 73F, cps	142
Phase coalescence temp., F	142
pH	11.6

Use Concentration:
Heavy-duty use: 4 oz/gal.
Regular-duty use: 2 oz/gal.

SOURCE: Shell Chemical Co.: NEODOL Formulary: Suggested
Formulation

FLOOR CLEANER/WAX STRIPPER

RAW MATERIALS   % By Weight

Water	86
Diethylene glycol butyl ether	4
PLURAFAC B-25-5 surfactant	4
Tetrapotassium pyrophosphate	6

Suggested use concentration: 2-4 oz per gallon of water

SOURCE: BASF Corp.: Cleaning Formulary: Formulation #3475

FLOOR CLEANER/WAX STRIPPER

RAW MATERIALS   % By Weight

DIACID H-240	2.3
Neodol 25-9	5.0
Trisodium phosphate	3.0
TKPP	5.0
Ammonium hydroxide	1.5
Water	q.s.*

*q.s.--quantity sufficient to make 100% total.

SOURCE: Westvaco Chemical Division: DIACID Surfactants:
        Suggested Formulation

WAX STRIPPER AND CLEANER

RAW MATERIALS   Parts By Weight

SURFONIC N-120	10
SURFONIC N-40	2
Tetrapotassium pyrophosphate*	2
Trisodium phosphate	1
Water	85

* Tetrasodium pyrophosphate may be substituted for tetrapotassium pyrophosphate

    If a slight haze is apparent in this formulation, the prod-
uct may be cleared by the addition of small quantities of
alcohol or by the addition of dyes which are usually used in
the commercial products. This product should be diluted 20 to 1
with water when used as a floor wax stripper. It is diluted 80
to 1 with water when used for general cleaning.

SOURCE: Texaco Chemical Co.: Suggested Formulation THAE2

LIQUID NON-PHOSPHATE FLOOR CLEANER AND WAX STRIPPER

RAW MATERIALS	% By Weight
NTA	3.0
Butyl cellosolve	4.0
NINOL 1281	10.0
Ammonium hydroxide (28%)	1.0
Monoethanolamine	5.0
Liquid potassium hydroxide	2-5
Silicate N	5.0
Water, (dye & perfume optional)	balance

Properties:
Appearance	clear yellow liquid
pH range	12.5-13.0
Viscosity, 25C, cps	10
% active	26-28.5

Use Instructions:
    Dilution ratio: 1-4 oz/gallon

Comments:
    Excellent for stripping of metallic crosslinked floor wax polymers in manual or automated scrubbing machines because of controlled foam, excellent rinseability and pick-up.

SOURCE: Stepan Co.: Formulation No. 37

FLOOR WAX STRIPPER

RAW MATERIALS	% By Weight
DOWANOL PM glycol ether	14.0
ammonium hydroxide (28%)	6.5
CARSONOL SHS	4.5
water	75.0

    DOWANOL PM provides penetration and is an excellent spot remover and wax stripper.

1. Mix chemicals in order listed.
2. For better mixing, melt the paraffin.
3. Ammonium hydroxide may be added for extra removal properties.

SOURCE: Dow Chemical Co.: The Glycol Ethers Handbook:
        Formulation II

EXPERIMENTAL STRIPPER

RAW MATERIALS	% By Weight	CAS Registry Number
Water	75.0	
Sodium Metasilicate	3.0	10213-79-3
Monoethanolamine	5.0	141-43-5
ESI-TERGE HA-20*	2.0	
ESI-TERGE 320*	4.0	52276-83-2
Butyl Cellosolve*	7.5	111-76-2
Isopropanol Alcohol*	2.5	67-63-0
Caustic Soda (NAOH 50%)	1.0	1310-73-2
	100.0	

Procedure:
    Use adequate agitation, add in order mentioned above.

    * Premix ESI-TERGE HA-20, ESI-TERGE 320, Butyl Cellosolve and Isopropanol Alcohol. Small amounts of defoamer may be used when vacuuming.

SOURCE: Emulsion Systems Inc.: Technical Service Bulletin
        Code PGR-14

HEAVY DUTY WAX STRIPPER

RAW MATERIALS	% By Weight	CAS Registry Number
Water	72.3	7732-18-5
Trisodium Phosphate	3.0	7601-54-9
Tetra Potassium Polyphosphate	3.0	7320-34-5
ESI-TERGE HA-20	5.0	Mixture
Monoethanolamine	5.0	141-43-5
Propasol BEP	10.0	
Sodium Xylene Sulfonate	1.7	1300-72-7

Procedure:
    Add salts to water and dissolve.
    Add other ingredients in order mentioned.

Specifications:
    % Solids   11.6
    % Active   26.6
    pH   11.5-12.5
    Viscosity   Water

SOURCE: Emulsion Systems Inc.: Technical Service Bulletin
        Code HA-20-9

FLOOR FINISH STRIPPER

RAW MATERIALS	% By Weight
Etnylene glycol n-butyl ether	10.0
TRYFAC 5556 Phosphate Ester	5.0
Triethanolamine (TEA)	3.0
TRYCOL 5941 POE (9) Tridecyl Alcohol	1.0
Tetrasodium EDTA (40%)	5.0
Soda Asn (sodium carbonate)	2.0
Sodium xylene sulfonate (40%) (SXS)	2.5
Dye and fragrance	as desired
Water	to 100

Blending Procedure:
    Add the water to the blending tank. While mixing, add the ingredients to the blending tank in the order listed. Stir until uniform.

SOURCE: Emery Chemicals: Specialty Chemicals Formulary: Formulation 2886-121

FOAMING WAX STRIPPER

RAW MATERIALS	% By Weight
AMPHOTERGE K	5.0
Sodium metasilicate pentahydrate	5.0
Tetrapotassium pyrophosphate	10.0
Water	80.0

SOURCE: Lonza Inc.: Product Information: Formulation M-2-1

WAX STRIPPER

RAW MATERIALS	% By Weight
MIRANOL J2M-SF CONC.	4.2
Dowanol EB	3.5
Isopropyl Alcohol	2.5
Potassium Hydroxide, 45%	4.0
Monoethanolamine	5.0
Tetrapotassium Pyrophosphate	2.0
Trisodium Phosphate	2.0
Water	76.8

SOURCE: Miranol, Inc.: MIRANOL Products for Household/Industrial Applications: Suggested Formulation

LOW FOAMING WAX STRIPPER

RAW MATERIALS	% By Weight
MIRANOL JEM CONC.	4.0
Tetrapotassium Pyrophosphate	4.8
Trisodium Phosphate	3.0
Starso	5.0
Dowanol EB	1.0
Water	82.2

SOURCE: Miranol Inc.: MIRANOL Products for Household/Industrial
        Applications: Suggested Formulation

LOW FOAM WAX STRIPPER

RAW MATERIALS	% By Weight
MIRAWET B	4.2
Trisodium Phosphate	2.0
Tetrapotassium Pyrophosphate	2.0
Monoethanolamine	5.0
Isopropyl Alcohol	2.5
Dowanol EB	3.5
Potassium Hydroxide, 45%	5.0
Water	75.8

SOURCE: Miranol Inc.: MIRANOL Products for Household/Industrial
        Applications: Suggested Formulation

<u>WAX STRIPPER</u>

RAW MATERIALS	% By Weight	CAS Registry Number
Water	86.00	7732-18-5
Trisodium Phosphate	3.00	7601-54-9
Sodium Tripolyphosphate	3.00	7758-29-4
ESI-TERGE HA-20	5.00	Mixture
Monoethanolamine	3.00	141-43-5

Procedure:
   Add in order listed with adequate agitation, allowing each material to dissolve before adding ESI-TERGE HA-20. Agitate 5 minutes.

Specifications:
% Solids	13.5
% Active	13.5
pH	10.5-11.5
Viscosity	Medium

SOURCE: Emulsion Systems Inc.: Technical Service Bulletin
        Code HA-20-8

<u>WAX STRIPPER</u>

RAW MATERIALS	% By Weight
MIRAWET FL	6.0
Tetrapotassium Pyrophosphate	10.0
Sodium Metasilicate Pentahydrate	6.0
Ammonium Hydroxide, 30%	8.0
Water	70.0

SOURCE: Miranol, Inc.: MIRANOL Products for Household/
        Industrial Applications: Suggested Formulation

<u>WAX STRIPPER</u>

RAW MATERIALS	% By Weight
MIRANOL J2M-SF CONC.	4.0
Monoethanolamine	3.0
Dowanol EB	5.0
Potassium Hydroxide, 45%	3.0
Tetrapotassium Pyrophosphate	3.0
Water	82.0

SOURCE: Miranol Inc.: MIRANOL Products for Household/Industrial
        Applications: Suggested Formulation

WAXED FLOOR CLEANER

RAW MATERIALS   % By Weight

GAFAC RA-600   5.0
GAFAMIDE CDD-518   1.0
Tetrapotassium pyrophosphate(60% active)   25.0
Water   69.0
                                                             100.0

Manufacturing Procedure:
    Add ingredients one-by-one, in order listed above, mixing
well after each addition.

Physical Properties
    pH (as is)   8.6
    pH (1%)   8.6
    Viscosity   10 cps
    Specific Gravity   1.04

SOURCE: GAF Corp.: Formulary: Prototype Formulation GAF 5351

HEAVY DUTY FLOOR FINISH STRIPPER

RAW MATERIALS   % By Weight

TRYFAC 5568 Phosphate Ester   5.0
Triethanolamine TEA   2.0
TRYCOL 6965 POE (11) Nonylphenol   2.0
EMERY 6705 Phenoxyethanol   5.0
Tetrasodium EDTA (40%)   5.0
Tetrapotassium pyrophosphate (TKPP)   5.0
Water   76.0

Blending Procedure:
    Add the water to the blending tank. While mixing, add the
remaining ingredients in the order listed. Be sure the TRYFAC
5568 has dissolved before adding the TKPP. Mix until uniform.

Use Dilution:
    Dilute 1 part of the formulated product to 5-25 parts of
hot water. The concentration will depend on the type of floor
finish and number of coats being stripped.

SOURCE: Emery Chemicals: Specialty Chemicals Formulary:
        Formulation 2886-121

5. General Purpose Cleaners

ABRASIVE CLEANER, LIQUID

RAW MATERIALS   % By Weight

Water   33.64
ACRYSOL ASE-108 (18%)   7.06
Sodium Hydroxide (10%)   4.30
Sodium Tripolyphosphate (STPP)   2.50
TRITON X-102   2.50
Berkeley 230 mesh (Jasper) or
 Berkeley 160 mesh Supersil   50.00
                                                               100.00

Properties:
    Solids Content, %   56.7
    Viscosity, cps/25C   9000
    Use Level   Use as made.

SOURCE: Rohm and Haas Co.: Specialty Chemicals: Detergent
        Formulations for Industrial and Institutional Industry:
        Ref: CS-407, CS-500

ABRASIVE CLEANER, LIQUID

RAW MATERIALS   % By Weight

Water   37.09
ACRYSOL ASE-108 Stabilizer   7.05
Sodium Hydroxide (50% solution)   0.86
Sodium Tripolyphosphate   2.50
TRITON X-102 Surfactant   2.50
Jasper Supersil or
 Extrafine Supersil   50.00

Properties:
    Solids Content, %   56.7
    Viscosity, cps/25C   9000
    Use Level   Use as made.

   Although abrasive household cleaners are readily available
in powdered form, similar products can be made as liquids by
taking advantage of the unique properties of ACRYSOL ASE-108
polymer. In this Formulation, the abrasive grit is kept in
suspension by the stabilizing action of ACRYSOL ASE-108 polymer.
The suspension is unaffected by multiple freeze-thaw cycles or
storage under normal conditions. The viscosity of the formulation is about 9,000 cps.

SOURCE: Rohm and Haas Co.: Specialty Chemicals: Detergent
        Formulations for Industrial and Institutional Industry:
        Lit. Ref: CS-407, CS-500

ABRASIVE CLEANER, LIQUID

RAW MATERIALS   % By Weight

Water	45.44
Calcium Carbonate	50.00
ACRYSOL ICS-1 Thickener (30%)	1.00
Bentonite Clay	2.50
TRITON X-100 Surfactant	0.50
Sodium Hydroxide (10%)	0.56
	100.00

Brookfield Viscosity, cps
   @ 0.5 rpm   18,000
   @ 12 rpm   5,500

Mixing Procedure:
   Add ingredients in stated order with moderate agitation.

SOURCE: Rohm and Haas Co.: Specialty Chemicals: Detergent
        Formulations for Industrial and Institutional Industry:
        Lit. Ref: CS-408/CS-427/CS-504: Cleaner A

ABRASIVE CLEANER, LIQUID

RAW MATERIALS   % By Weight

Water	43.40
Silica	50.00
ACRYSOL ICS-1 Thickener (30%)	1.50
Tetrapotassium Pyrophosphate (50%)	4.00
TRITON N-101 Surfactant	0.25
Sodium Hydroxide (10%)	0.85
	100.00

Brookfield Viscosity, cps
   @ 0.5 rpm   53,000
   @ 12 rpm   12,400

Mixing Procedure:
   Add ingredients in stated order with moderate agitation.

SOURCE: Rohm and Haas Co.: Specialty Cleaners: Detergent
        Formulations for Industrial and Institutional Industry:
        Lit. Ref: CS-408/CS-427/CS-504: Cleaner B

ALKALINE ALL PURPOSE INDUSTRIAL CLEANER
Phosphated, Powder Type

RAW MATERIALS	% By Weight
Sodium tripolyphosphate (lt. density)	52.5
Sodium carbonate (lt. density)	20.0
Sodium metasilicate 5-H2O	22.5
GAFAC RA-600	5.0
	100.0

Manufacturing Procedure:
1. Mix GAFAC RA-600 with sodium tripolyphosphate to obtain a uniform powdered mixture.
2. Add sodium metasilicate 5-H2O and sodium carbonate.

Physical Properties:
pH (1%)	11.4
Specific Gravity	.82

SOURCE: GAF Corp.: Formulary: Prototype Formulation GAF 5601

ALL-SURFACE HOUSEHOLD CLEANER

RAW MATERIALS	% By Weight
Tetrasodium EDTA (40%)	1.0
TRYCOL 5941 POE (9) Tridecyl Alcohol	1.0
EMID 6533 Modified Alkanolamide	4.0
TRYCOL 6964 POE (9) Nonylphenol	4.0
Triethanolamine (TEA)	0.5
Dye, Fragrance, etc.	as desired
Water	to 100

pH (as is) = 9.3

Blending Procedure:
Add the water to the blending tank. While mixing, add the ingredients to the blending tank in the order listed. Stir until uniform.

Use Dilution:
2 to 4 ounces (1/4 to 1/2 cups) per gallon of water.

SOURCE: Emery Chemicals: Specialty Chemicals Formulary: Formulation No. 2749-005

ALL PURPOSE CLEANER

RAW MATERIALS   % By Weight

MAZER MAPHOS 91   5
MAZER MAZON 60T   5
MAZER MAZAMIDE 70   1
Sodium Tripolyphosphate   10
Water   79

SOURCE: Mazer Chemicals, Inc.: Household/Industrial T-20B: 2

ALL PURPOSE CLEANER TYPE (AJAX)

RAW MATERIALS   % By Weight

REWORYL TKS 90/F   3.0
REWOPOL HV 10   2.0
Urea   3.5
Ammonia (.880)   2.0
EDTA   0.4
Potassium Soap 50%   0.7
Sodium Carbonate   15.0
REWORYL NXS 40   8.0
water   qs 100.0

Mixing Procedure:
   Dissolve the Carbonate and Urea in water. Stir in the other
components.

SOURCE: Sherex Chemical: Industrial Formulation 52:01.6

ALL PURPOSE CLEANER

RAW MATERIALS   % By Weight

HOSTAPUR SAS 60   10.0
Isotridecylalcoholethoxilate (8EO)   3.0
Tetrapotassium Pyrophosphate   5.0
Ammonia, 25%   1.0
Isopropyl Alcohol   5.0
Water, Perfume Oil and Dye   AD 100.0

SOURCE: Hoechst/Celanese: Suggested Formulation

ALL PURPOSE CLEANER
Transparent, Liquid

RAW MATERIALS   % By Weight

HOSTAPUR SAS 60   7.0
Nonylphenolethoxilate (8EO)   2.5
Sodium Citrate   6.0
Na2CO3   2.0
Perfume   0.2
Water, Preservative, Dye   82.3

Production Procedure:
  Dissolve sodium citrate and sodium carbonate in water and
mix it with HOSTAPUR SAS, Nonionic, Perfume, Preservative and
Dye.

Tests:
    pH Value (10% Aqueous Solution T.Q.)   8.7
    Viscosity (Brookfield RVT)   55 MPAS
    Freeze and Thaw Test   O.K.

SOURCE: Hoechst/Celanese: Formulation D-1012

ALL PURPOSE CLEANER

RAW MATERIALS   % By Weight

TRYCOL 6964 POE (9) Nonylphenol   15.0
TRYCOL 5941 POE (9) Tridecyl Alcohol   20.0
TRYCOL 5951 POE (5) Decyl Alcohol   2.0
Tetrasodium EDTA (40%)   2.5
Triethanolamine (TEA)   0.5
Dye, fragrance, etc.   as desired
Water   to 100

Blending Procedure:
  Add the water to the blending tank. While mixing, add the
ingredients to the blending tank in the order listed. Stir
until uniform.

Use Dilution:
  1 to 2 ounces (1/8 to 1/4 cup) per gallon of water.

SOURCE: Emery Chemicals: Specialty Chemicals Formulary:
        Formulation 2878-013(2D)

ALL PURPOSE CLEANER

RAW MATERIALS	% By Weight
TRYCOL 6964 POE (9) Nonylphenol	20.0
TRYCOL 5941 POE (9) Tridecyl Alcohol	30.0
TRYCOL 5951 POE (5) Decyl Alcohol	2.0
Tetrasodium EDTA (40%)	2.5
Triethanolamine (TEA)	0.5
Dye, fragrance, etc.	as desired
Water	to 100

Blending Procedure:
   Add the water to the blending tank. While mixing, add the ingredients to the blending tank in the order listed. Stir until uniform.

Use Dilution:
   1 to 2 ounces (1/8 to 1/4 cup) per gallon of water.

SOURCE: Emery Chemicals: Specialty Chemicals Formulary: Formulation No. 2878-013(3D)

ALL PURPOSE CLEANER

RAW MATERIALS	% By Weight	CAS Registry Number
Water	89.0	7732-28-5
Trisodium Phosphate	3.0	7601-54-9
Sodium Tripolyphosphate	3.0	7758-29-4
ESI-TERGE HA-20	5.0	Mixture
	100.0	

Procedure:
   Add salts to water and dissolve. Add other ingredients in order mentioned.

Specifications:
% Solids	11.0
% Active	11.0
pH	10-11
Viscosity	Medium

SOURCE: Emulsion Systems Inc.: Technical Service Bulletin Code HA-20-1

ALL PURPOSE CLEANER

RAW MATERIALS   % By Weight

MIRANOL C2M-SF CONC.   10.0
Tetrapotassium Pyrophosphate   5.0
Tetrasodium Pyrophosphate   5.0
Potsssium Hydroxide, 45%   4.0
Actinol FA-2   10.0
Water   66.0

Procedure:
    Dissolve the TKPP and TSPP at 60-70C, then add the potassium
hydroxide liquid. With stirring and heating add and disperse the
Actinol FA-2. Finally, add MIRANOL C2M-SF CONC. When uniform,
allow to stand until clear.

Note:
    The appearance of a white cream which this product develops
during mixing will disappear after the foam and air have risen.
The end product will be a crystal clear liquid having a high
viscosity and a pH of 9.8-10.2.

SOURCE: Miranol Chemical Co.: MIRANOL Products for Household/
        Industrial Applications: Suggested Formulation

ALL PURPOSE CLEANER(AEROSOL)

RAW MATERIALS   % By Weight

AEROTHENE TT solvent   30.00
AEROTHENE MM solvent   7.50
DOWANOL DPM glycol ether   6.00
toluene   22.50
diisopropanolamine   2.25
TRITON X-100 surfactant   1.50
amyl acetate   5.25
propellant A-70   25.00

Suggested Valve: Seaquist .012" stem/.018" body
Suggested Actuator: Seaquist RKN-15

    DOWANOL DPM is a good solvent for inks. Toluene and AEROTHENE
solvents are good grease, oil and wax solvents.

SOURCE: Dow Chemical U.S.A.: The Glycol Ethers Handbook:
        Suggested Formulation

ALL PURPOSE CLEANER WITH FBR

RAW MATERIALS	% By Weight
REWORYL NKS 50	4.0
VARAMIDE FBR	6.0
VAROX 365	6.0
REWORYL NXS 40	6.0
Sodium Tripolyphosphate (STP)	2.5
Water	75.5

Mixing Procedure:
   Dissolve STP in water. Stir in VAROX 365. Stir NXS 40 into the solution. Stir in NXS 50 and FBR into the solution. Let sit until air bubbles settle. Final product should be clear with gold color.

SOURCE: Sherex Chemical: Industrial Formulation 23:1.6

ALL PURPOSE
(WALL/TILE/FLOOR CLEANER)

RAW MATERIALS	% By Weight
NEODOL 91-6	6.0
FADEA	7.0
Trisodium phosphate, anhydrous basis	2.0
Sodium metasilicate, pentahydrate	13.9
Isopropyl alcohol	2.0
Water, dye, perfume	to 100%

Properties:
Viscosity, 73F, cps	27
Phase coalescence temp., F	106
pH	13.0

SOURCE: Shell Chemical Co.: The NEODOL Formulary: Suggested
        Formulation

ALL PURPOSE JANITORIAL CLEANING CONCENTRATE

RAW MATERIALS	% By Weight
MIRANOL C2M-SF CONC.	10.0
Tetrapotassium Pyrophosphate	30.0
Starso	16.0
Water	44.0

Note: This formulation is to be used as a concentrate, 1/4 to
      1/2 ounce in 10-12 quarts of water. It may be used for
      floors, walls, painted surfaces and appliances.

SOURCE: Miranol Chemical Co.: MIRANOL Products for Household/
        Industrial Applications: Suggested Formulation

ALL-PURPOSE CONCENTRATE CLEANER

RAW MATERIALS   % By Weight

TRITON QS-44 Surfactant (80%)   5.0
Sodium Hydroxide   0.9
TRITON X-100 Surfactant   4.0
Tetrapotassium Pyrophosphate (TKPP)   16.0
Water   74.1
                                                              100.0

Use Directions: For heavy soiling 1 part/4 parts water
                For most cleaning 1 part/64 parts water

SOURCE: Rohm and Haas Co.: Specialty Chemicals: Detergent
        Formulations for Industrial and Institutional Industry:
        Lit. Ref: CS-410/CS-427

ALL-PURPOSE CONCENTRATE CLEANER

RAW MATERIALS   % By Weight

Water   72.7
ACRYSOL ASE-108 Stabilizer   4.4
Sodium Hydroxide (10%)   2.6
Tetrapotassium Pyropnosphate (anhydrous)   5.0
Tetrasodium Pyrophosphate (anhydrous)   5.0
TRITON X-100 Surfactant   8.0
Lauricdiethanolamide (87%)   2.3
Dye   0.002

Mixing Instructions:
   Add ingredients in listed order with vigorous subsurface
agitation. Perfumes and colorants may be used at the formulator's
discretion.

Properties:
    Appearance   Opaque low-viscosity liquid
    Solids, %   21.2
    pH (1.5% aqueous solution) 9.9
    Density, lb./gal.   9.2
    Specific Gravity @ 25C.   1.1
    Viscosity, cps @ 25C.   350

Use Dilution:
    1/4 cup per gallon of water for general household cleaning.

SOURCE: Rohm and Haas Co.: Specialty Chemicals: Detergent
        Formulations for Industrial and Institutional Industry:
        Lit. Ref: CS-427, CS-500

ALL-PURPOSE CONCENTRATE CLEANER

RAW MATERIALS	% By Weight
TRITON QS-30 Surfactant (90%)	3.3
TRITON X-100 Surfactant	6.7
Tetrapotassium pyrophosphate (TKPP)	10.0
Water	80.0
	100.0

Use Directions:
   For heavy soils--1 part/4 parts water
   For most cleaning--1 part/64 parts water

SOURCE: Rohm and Haas Co.: Specialty Chemicals: Detergent
        Formulations for Industrial and Institutional Industry:
        Lit. Ref.: CS-427/CS-439

ALL-PURPOSE CLEANER(LIQUID CONCENTRATE)
(Clear Heavy-Duty Concentrate)

RAW MATERIALS	% By Weight
TRITON X-100 Surfactant	25.00
TRITON X-45 Surfactant	10.00
Diethanolamine	10.00
Ethanol 3A	7.60
FD & C Blue #1	0.05
Tergescent No. 7	0.10
Water	47.25
	100.00

Use Dilution: 1 oz/2 gallons water.

SOURCE: Rohm and Haas Co.: Specialty Chemicals: Detergent
        Formulations for Industrial and Institutional Industry:
        Lit. Ref.: CS-403/CS-427

ALL-PURPOSE LIQUID CONCENTRATE CLEANER-A

RAW MATERIALS	% By Weight
TRITON QS-44 Surfactant(80%)	3.00
Sodium Hydroxide	0.55
TRITON X-100 Surfactant	7.60
Tetrasodium pyrophosphate	5.00
Tetrapotassium pyrophosphate	5.00
Water	78.85
	100.00

Use Dilution: 3 to 6 oz./gal. water

SOURCE: Rohm and Haas Co.: Specialty Chemicals: Detergent
        Formulations for Industrial and Institutional Industry:
        Lit. Ref: CS-410, CS-427

ALL PURPOSE HEAVY DUTY CLEANER

RAW MATERIALS	% By Weight
AMPHOTERGE K	10.0
Sodium metasilicate pentanydrate	15.0
Sodium carbonate	5.0
Ethylene diamine tetraacetic acid tetrasodium salt (38% sol'n)	5.0
Ammonium hydroxide (28% sol'n)	3.5
Water	61.5

SOURCE: Lonza: Product Information Amphoterge K/K-2:
        Formulation N-107-2

ALL PURPOSE LIQUID INDUSTRIAL CLEANER

RAW MATERIAL	% By Weight
MIRANOL C2M-SF CONC.	4.0
CEDEMIDE CX	4.5
Tetrapotassium Pyrophosphate	2.4
Trisodium Nitrilotriacetate, 40% solution	3.5
Potassium Hydroxide, 45%	1.6
Triton X-100	4.0
Water	80.0

SOURCE: Miranol Inc.: MIRANOL Products for Household/Industrial
        Applications: Suggested Formulation

ALL-PURPOSE LIQUID DETERGENT-A

RAW MATERIALS	% By Weight
TRITON X-102 Surfactant	5.0
TRITON H-66 Surfactant (50%)	2.0
Tetrapotassium Pyrophosphate (TKPP)	5.0
Dipropylene Glycol Methyl Ether (Dowanol DPM)	10.0
Tetrasodium Ethylenediaminetetraacetate (Versene 100)	3.0
Water	75.0
	100.0

Use Dilution: 1 part/16 parts water

SOURCE: Rohm and Haas Co.: Specialty Chemicals: Detergent
        Formulations for Industrial and Institutional Industry:
        Lit. Ref.: CS-407/CS-433

ALL PURPOSE HOUSEHOLD DETERGENT

RAW MATERIALS   % By Weight

Ninol 1285 surfactant   3.0
sodium xylene sulfonate   6.0
trisodium phosphate   5.0
DOWANOL DPM glycol ether   4.0
pine oil   2.0
water   80.0

1. Dissolve the surfactants and TSP in the water.
2. Add DOWANOL and pine oil next.
3. Mix well.

SOURCE: Dow Chemical U.S.A.: The Glycol Ethers Handbook:
        Suggested Formulation

ALL PURPOSE INDUSTRIAL CLEANER
Phosphated, Moderate Duty Type

RAW MATERIALS   % By Weight

GAFAC RA-600   5.0
Alkylbenzene sulfonic acid   5.0
Sodium hydroxide (50% active)   1.3
GAFAMIDE CDD-518   1.0
Sodium tripolyphosphate (lt. density)   10.0
Water   77.7
                                                            100.0

   Perfume and colorants added, as desired, replacing water.

Manufacturing Procedure:
   1. Dissolve alkylbenzene sulfonic acid in one half total
      amount of water.
   2. Neutralize with sodium hydroxide.
   3. Add GAFAC RA-600 and GAFAMIDE CDD-518 individually,
      mixing well after each addition.
   4. Dissolve sodium tripolyphosphate in remaining amount
      of water. Add to main batch.

Physical Properties:
   pH (as is)   7.4
   pH (1%)   11.4
   Viscosity   160 cps
   Specific Gravity   1.03

SOURCE: GAF Corp.: Formulary: Prototype Formulation GAF 5651

ALL-PURPOSE LIGHT DUTY LIQUID DETERGENT

RAW MATERIALS	% By Weight
TRITON X-301 Surfactant (20%)	50.0
TRITON X-100 Surfactant	20.0
Water	30.0
	100.0

Perfumes and dyes may be added. The above basic formulation can be modified to provide floor cleaners, car washes, industrial and household hand dishwashing compounds, and fine fabric detergents.

SOURCE: Rohm and Haas Co.: Specialty Chemicals: Detergent
        Formulations for Industrial and Institutional Industry:
        Lit. Ref: CS-33, CS-427

ALL-PURPOSE CLEANER, LIGHT DUTY

RAW MATERIALS	% By Weight
TRITON X-100 Surfactant	8.0
TRITON H-66 Surfactant (50%)	4.0
Tetrapotassium Pyrophosphate	10.0
Water	78.0
	100.0

Use Dilution: 3 oz/gal. water.

SOURCE: Rohm and Haas Co.: Specialty Chemicals: Detergent
        Formulations for Industrial and Institutional Industry:
        Lit. Ref: CS-427, CS-433

ALL PURPOSE CLEANER

RAW MATERIALS	% By Weight
VARION AMKSF40	10.0
Tall Oil Fatty Acid	10.0
45% Potassium Hydroxide	3.0
Potassium Pyrophosphate	4.0
Tetra Potassium Pyrophosphate	4.0
Water	qs100

Mixing Procedure:
    Dissolve the Phosphates in the water. Then add the potassium hydroxide followed by the Tall Oil Fatty Acid to the required ph. Finally add the AMKSF 40

SOURCE: Sherex Chemical: Industrial Formulation 43:0.16

ALL-PURPOSE LIQUID CONCENTRATE CLEANER

RAW MATERIALS	% By Weight
TRITON QS-44 Surfactant (80%)	5.00
Sodium Hydroxide	0.90
TRITON X-100 Surfactant	4.00
Tetrapotassium pyrophosphate	16.00
Water	74.10
	100.00

Use Dilution: 3 to 6 oz./gal. water

SOURCE: Rohm and Haas Co.: Specialty Chemicals: Detergent
Formulations for Industrial and Institutional Industry:
Lit. Ref.: CS-410, CS-427

ALL-PURPOSE CLEANER, LIQUID

RAW MATERIALS	% By Weight
Water	74.68
ACRYSOL ASE-95 Thickener (18%)	4.72
Sodium Hydroxide (50%)	0.60
Tetrapotassium Pyropnosphate (Granular)	10.00
TRITON X-100 Surfactant	10.00
	100.00

Properties:
Solids Content, %	21.15
Density, lb./gal.	9.2
Viscosity, cps/25C	2100
Use Level	1/4 cup per gallon of water for general household cleaning

SOURCE: Rohm and Haas Co.: Specialty Chemicals: Detergent
Formulations for Industrial and Institutional Industry:
Lit. Ref: CS-427/CS-501

ALL PURPOSE CLEANER

RAW MATERIALS	% By Weight
MAZER MAZON 41	25.0
MAZER MACOL 48	5.0
Pine Oil (steam distilled)	1.0
MAZER MAZON 71A	1.0
Versene	1.0
Water	67.0

SOURCE: Mazer Chemicals, Inc.: Household/Industrial T-20B: 3

ALL-PURPOSE LIQUID DETERGENT

RAW MATERIALS	% By Weight
TRITON X-102 Surfactant	5.0
TRITON H-66 Surfactant (50%)	15.0
Tetrapotassium Pyrophosphate (TKPP)	15.0
Dipropylene Glycol Methyl Ether (Dowanol DPM)	4.0
Tetrasodium Ethylenediaminetetraacetate (Versene 100)	3.0
Water	58.0

Use Dilution: 1 part/16 parts water

SOURCE: Rohm and Haas Co.: Specialty Chemicals: Detergent
Formulations for Industrial and Institutional Industry:
Lit. Ref: CS-407/CS-433

ALL-PURPOSE LIQUID DETERGENT

RAW MATERIALS	% By Weight
Water	77.0
Ivory Flakes	5.0
Tetrapotassium Pyrophosphate (TKPP)	10.0
TRITON X-102 Surfactant	5.0
Propylene glycol	3.0
	100.0

Mixing Instructions:
   Heating may be needed to dissolve the soap flakes.
   Add the other ingredients after the soap dissolves.

Appearance:
   Clear slightly yellow, slightly viscous liquid.

Use Dilutions: Heavy work--1 part/4 parts water
               Most work--1 part/64 parts water

Note: White vinyl floor tiles may be discolored by misuse of
      cleaners containing a combination of potassium salt
      builder and nonionic detergent. Avoid excessive concentra-
      tion and long contact.

SOURCE: Rohm and Haas Co.: Specialty Chemicals: Detergent
Formulations for Industrial and Institutional Industry:
Lit. Ref.: CS-407

ALL-PURPOSE LIQUID DETERGENT

RAW MATERIALS   % By Weight

TRITON X-100 Surfactant	10.00
Alkylarylsulfonic acid (98%)	2.50
Borax	2.00
Tetrapotassium Pyrophosphate (TKPP)	2.00
Water	82.65
Sodium Hydroxide (50%)	.85
	100.00

Use Dilutions: General Use--1 part/64 parts water
                Heavy-duty--1 part/4 parts water

SOURCE: Rohm and Haas Co.: Specialty Chemicals: Detergent
         Formulations for Industrial and Institutional Industry:
         Lit. Ref: CS-427

ALL-PURPOSE LIQUID PINE OIL DETERGENT

RAW MATERIALS   % By Weight

TRITON X-100 Surfactant	5.00
TRITON X-45 Surfactant	2.50
Dipropylene Glycol Methyl Ether (Dowanol DPM)	6.00
Pine Oil	0.25
Tetrapotassium Pyrophosphate (TKPP)	3.00
Sodium Metasilicate Pentahydrate	2.00
TRITON H-66 Surfactant (50%)	2.00
Water	79.25
	100.00

Appearance:
   Cloudy as prepared. At 1/64 dilution clear.

Use Dilutions: Most work--1/64
                Heavy work--1/4
                Tough work--as prepared

SOURCE: Rohm and Haas Co.: Specialty Chemicals: Detergent
         Formulations for Industrial and Institutional Industry:
         Lit. Ref: CS-403/CS-427/CS-433

ALL-PURPOSE SPRAY CLEANER

RAW MATERIALS	% By Weight
EMID 6538 Modified Alkanolamide	3.0
TRYCOL 6964 POE (9) Nonylphenol	2.0
Ethylene glycol n-butyl ether	5.0
Tetrasodium EDTA (40%)	2.0
Sodium hydroxide (50%) (to pH 11.2-11.7)	q.s.
Dye, fragrance, etc.	as desired
Water	to 100

Blending Procedure:
  Add the water to the blending tank. While mixing, add the remaining ingredients in the order listed

Use Direction:
    The above formula is used "as is" and packaged in either a pump spray container or aerosol. Assure compatibility with packaging materials.

SOURCE: Emery Chemicals: Specialty Chemicals Formulary: Formulation 2749-006

ALL PURPOSE CLEANER WITH ALKANOLAMIDES

RAW MATERIALS	% By Weight
Sodium Lauryl Ether Sulfate 28% (SLES)	42.9
VARAMIDE 6CM	5.1
VARAMIDE A10	4.1
Tetrapotassium Pyrophosphate	5.1
Water	42.8

Mixing Procedure:
    Stir in SLES to water. Dissolve Tetrapotassium Pyrophosphate into the solution. Stir in VARAMIDE 6CM and then VARAMIDE A10

SOURCE: Sherex Chemical: Industrial Formulation 22:1.6

CAUSTIC CLEANER

RAW MATERIALS	% By Weight
MIRAPON JAS-50	3.0
Potassium Hydroxide (45%)	10.0
Kasil #6	50.0
Water	37.0

SOURCE: Miranol Inc.: MIRANOL Products for Household/Industrial
Applications: Suggested Formulation

CHLORINE CLEANER(KITCHEN AND BATHROOM)(POWDER)

RAW MATERIALS	% By Weight
PHOSPHAT SPR II	30.0
HOSTAPUR SAS 60	16.0
Silicat (5'er Hydrat)	30.0
Na2SO4	18.5
Sodiumdichloroisocyanurat	5.0
Aerosil 200	2.0

Production procedure:
   PHOSPHAT SPR II and HOSTAPUR SAS 60 will be mixed and Silicat,
Na2SO4 and Aerosil 200 should be added slowly. The mixture should
be dried at 50 - 60C and afterwards pulverized. Sodiumdichlor-
isocyanurat will then be added and distributed in the mixture.
Tests:
   flow density   1065
   flowability (DIN 53916)   1.78 cot
   pH-value (10%)   11.9

SOURCE: Hoechst/Celanese: Formulation D-6017

FOOD INDUSTRY CLEANER-GENERAL USE LIQUID

RAW MATERIALS	% By Weight
NEODOL 23-6.5	5.0
Sodium metasilicate, pentahydrate	5.0
Tetrapotassium pyrophosphate	3.0
Triton H-66	5.0
Butyl OXITOL	5.0
Water, dye, perfume	to 100%

Properties:
   Viscosity, 73F, cps   8
   Phase coalescence temp., F   135
   pH   13.4
Use Concentration: 1-2 oz/gal

SOURCE: Shell Chemical Co.: NEODOL Formulary: Suggested Formula

FOOD INDUSTRY CLEANER-POWDER

RAW MATERIALS	% By Weight
NEODOL 23-6.5	5.0
Sodium carbonate	36.5
Sodium hydroxide, flakes	21.3
Sodium metasilicate, pentanydrate	18.6
Tetrasodium pyrophosphate	23.6

Blending Procedure for Powder Only:
   Mix solid builders thoroughly. Add surfactant slowly with mixing, mix thoroughly.

SOURCE: Shell Chemical Co.: NEODOL Formulary: Suggested Formula

GENERAL CLEANER--ALKALINE CLEANER

RAW MATERIALS	% By Weight
AEROSOL OS	5
Sodium Metasilicate	50
Caustic Soda	5
Tetrasodium Pyrophosphate	7
Trisodium Phosphate	33

SOURCE: Angus Chemical Co.: Suggested Formulation

GENERAL CLEANER-HOUSEHOLD

RAW MATERIALS	% By Weight
Tetrapotassium Pyrophosphate	8.0
Dodecylbenzene Sulfonate Triethanolamine Salt	4.0
AEROSOL A-103	4.0
Lauric Diethanolamide	2.0
Sodium Xylene Sulfonate	6.0
Water	76.0

SOURCE: Angus Chemical Co.: Suggested Formulation

GENERAL PURPOSE CLEANER--HOUSEHOLD

RAW MATERIALS   % By Weight

Tetrapotassium pyrophosphate   8
Dodecylbenzene sulfonate, TEA salt   4
SURFONIC N-95   4
Lauric diethanolamide   2
Sodium xylene sulfonate   6
Water   76

SOURCE: Texaco Chemical Co.: SURFONIC N-Series Surface-Active
        Agents: Suggested Formulation

GENERAL PURPOSE CLEANER-INDUSTRIAL

RAW MATERIALS   % By Weight

Tetrapotassium pyrophosphate   4
Dodecylbenzene sulfonate, TEA salt   4
SURFONIC N-150   5
Lauric diethanolamide   1
Water   86

SOURCE: Texaco Chemical Co.: SURFONIC N-Series Surface-Active
        Agents: Suggested Formulation

GENERAL PURPOSE CLEANER

RAW MATERIALS   % By Weight

Nonionic(nonylphenol-10EO)   8
Sodium tripolyphosphate   5
Trisodium phosphate   5
ACTRASOL SR606   3
Water   79

   If desired, coco alkanolamides can be substituted for the
nonylphenol ethoxylate, as is often done for floor cleaners.

SOURCE: Arthur C. Trask Corp.: The ACTRASOLS: Suggested
        Formulation

GENERAL PURPOSE SPRAY AND WIPE CLEANER

RAW MATERIALS	% By Weight
NINOL 1281	1.5
NTA	2.0
Sodium metasilicate, anhydrous	1.0
Butyl cellosolve	3.0
Water, color, perfume	Balance

Use Instructions:
    Can be used as is in trigger spray bottle

Comments:
    Similar in function to FANTASTIK. Leaves surfaces clean and bright.

SOURCE: Stepan Co.: Formulation No. 41

GENERAL PURPOSE SPRAY AND WIPE CLEANER

RAW MATERIALS	% By Weight
NINOL 11-CM	1.5
Na3NTA	2.0
Sodium metasilicate, anhydrous	1.0
Butyl cellosolve	3.0
Water, color, perfume	balance

Mixing Procedure:
    Dissolve NTA and sodium metasilicate in water, add butyl cellosolve and NINOL 11-CM in that order and mix.

Properties:
    Appearance   Clear liquid
    Butyl cellosolve odor

Use Instructions:
    Use as is from trigger spray bottle.

Performance:
    Leaves surface clean and bright.

Comments: Similar in function to FANTASTIK.

SOURCE: Stepan Co.: Formulation No. 85

GENERAL PURPOSE SPRAY AND WIPE CLEANER

RAW MATERIALS	% By Weight
Water, deionized	90.5
STEPANATE X	2.0
Na4EDTA	2.0
Sodium metasilicate, anhydrous	1.0
Butyl cellosolve	3.0
BIO SOFT LD-190	1.5

Mixing Procedure: Blend ingredients in order given.

Properties:
Appearance	clear liquid
Viscosity @ 25C, cps	4
pH as is	12.9
Specific Gravity	1.02
Density, lbs/gal	8.47

Use Instructions:
   Use as is in trigger spray bottle. Spray on and wipe off with clean cloth or paper towel.

SOURCE: Stepan Co.: Formulation No. 47

GENERAL PURPOSE CLEANER
Metal Parts Cleaner

RAW MATERIALS	% By Weight
pine oil	62.0
oleic acid	10.8
triethanolamine	7.2
DOWANOL DPM glycol ether	10.8
DOWANOL DB glycol ether	10.0

Dilute with equal volume of naphtha and white spirits. Water is added slowly with agitation.
For soil removal (not a rust remover).
DOWANOL used as a penetrant and coupling solvent.

SOURCE: Dow Chemical U.S.A.: The Glycol Ethers Handbook: Suggested Formulation

HEAVY DUTY ALKALINE CLEANER

RAW MATERIALS	% By Weight
MIRATAINE H2C	4.0
Sodium Gluconate	5.0
Sodium Hydroxide, 50%	44.0
Water	47.0

SOURCE: Miranol Inc.: MIRANOL Products for Household/Industrial Applications: Suggested Formulation

HEAVY DUTY CLEANER

RAW MATERIALS	% By Weight
MIRATAINE H2C	5.0
Sodium Metasilicate Pentahydrate	3.0
Trisodium Phosphate	1.3
Sodium Tripolyphosphate	1.5
Actinol FA-2	1.7
Potassium Hydroxide, 45%	1.0
Dowanol EB	9.0
Water	77.5

SOURCE: Miranol Inc.: MIRANOL Products for Household/Industrial Applications: Suggested Formulation

HEAVY DUTY, GOOD QUALITY, ALL PURPOSE SPRAY
WALL/TILE/FLOOR CLEANER

RAW MATERIALS	% By Weight
NEODOL 23-6.5	2.4
EDTA	2.6
Butyl OXITOL	3.0
Isopropyl Alcohol	1.0
Water, dye, perfume	to 100%

Properties:

Viscosity, 73F, cps	6
Phase coalescence temp., F	137
pH	10.9

SOURCE: Shell Chemical Co.: NEODOL Formulary: Suggested Formula

HEAVY DUTY CLEANER*

RAW MATERIALS	% by Weight	CAS REGISTRY NUMBER
Water	84.0	
Trisodium Phosphate	5.0	7601-54-9
Sodium Metasilicate	5.0	10213-79-3
ESI-TERGE 320	4.0	52276-83-2
ESI-TERGE DDBSA	2.0	27176-87-0
	100.0	

Procedure:
   Dissolve trisodium phosphate and sodium metasilicate in water. Add ESI-TERGE 320 and ESI-TERGE Dodecyl Benzene Sulfonic Acid. Agitate until clear.

Specifications:

% Solids	16
% Activity	16
pH	12-13
Viscosity	Medium

* To convert to a wax stripper, 3-5% ammonia or monoethanol-amine is added to this cleaner.

SOURCE: Emulsion Systems Inc.: Technical Service Bulletin Code
        320-6

HEAVY DUTY ALL PURPOSE STEAM CLEANER

RAW MATERIALS	% By Weight	CAS REGISTRY NUMBER
Water	81.8	7732-18-5
Sodium Metasilicate	8.7	10213-79-3
Caustic Soda	4.3	1310-73-2
Trisodium Phosphate	2.2	7601-54-9
ESI-TERGE 320	3.0	52276-83-2
	100.0	

Procedure:
   Add salts to water and dissolve. Add other ingredients in order mentioned.

Specifications:
% Solids	14.3
% Active	14.3
pH	14
Viscosity	3 cps.

SOURCE: Emulsion Systems Inc.: Technical Service Bulletin Code 320-12

INDUSTRIAL CLEANER

RAW MATERIALS	% By Weight
MIRAWET FL	10.0
Potassium Hydroxide, 45%	10.0
Versene 100	5.0
Kasil #1	25.0
Water	50.0

SOURCE: Miranol Inc.: MIRANOL Products for Household/Industrial Applications: Suggested Formulation

LIGHT DUTY CLEANER

RAW MATERIALS	% By Weight
Varamide A 10	10.0
Potassium Pyrophosphate	5.0
GAFAC RA-600	2.0
Water	qs100

Mixing Procedure: Add into water the order shown

SOURCE: Sherex: Industrial Formulation 16:01.6

LIGHT DUTY, ALL PURPOSE
WALL/TILE/FLOOR CLEANER

RAW MATERIALS   % By Weight

NEODOL 91-6   5.0
Trisodium phosphate, anhydrous basis   2.0
Sodium metasilicate, pentahydrate   3.5
Water, dye, perfume   to 100%

Properties:
  Viscosity, 73F, cps   8
  Phase coalescence temp., F   106
  pH   12.3

SOURCE: Shell Chemical Co.: NEODOL Formulary: Suggested
        Formulation

LIQUID CONCENTRATE

RAW MATERIALS   % By Weight

DIACID H-240   1.4
Sodium mestsilicate   1.0
Neodol 25-9   5.0
TKPP   10.0
Water   q.s.*

    *q.s.--quantity sufficient to make 100% total

SOURCE: Westvaco Chemical Division: DIACID Surfactants:
        Suggested Formulation

LIQUID HIGH PRESSURE CLEANER(CONCENTRATE)

RAW MATERIALS   % By Weight

VARION AMKSF 40   20.0
Tetrapotassium Pyrophosphate (TKPP)   20.0
Sodium Metasilicate Pentahydrate (SMSP)   10.0
Sodium Benzoate (Na Benzoate)   1.0
Sodium Xylene Sulfonate (SXS)   1.0
Dowanol EM   2.0
Water   qs 100

Mixing Procedure:
    Dissolve the EM, SXS, SMSP, TKPP and the Na Benzoate into
water. Now add the AMKSF 40.

SOURCE: Sherex: Industrial Formulation 4:05.6

LIMONENE BASED ALL PURPOSE CLEANER
CLEAR LIQUID

RAW MATERIALS	% By Weight
d-Limonene	50.0
NINATE 411	12.5
MAKON 4	7.5
MAKON 10	5.0
Water, soft	25.0

Mixing Procedure:
    Add surfactants to d-Limonene, mix until clear. Add water slowly while under continouus high agitation. Mix until clear.
Properties:
Appearance	Clear liquid
Odor	Citrus
Viscosity, cps @ 25C	20

Use Instructions:
    Use as is or dilute with water.
Performance:
    Removes grease, tar, chewing gum and most other oily soils effectively.

SOURCE: Stepan Co.: Formulation No. 110

LIMONENE BASED HEAVY DUTY CLEANER/DEGREASER

RAW MATERIALS	% By Weight
d-Limonene	50.0
NINOL 11-CM	9.5
MAKON 12	5.0
AMMONYX LO	0.5
Butyl cellosolve	10.0
Water, soft	25.0

Mixing Procedure:
    Add surfactants and Butyl cellosolve to d-Limonene, mix until clear. Add water slowly while under continuous high agitation. Mix until clear.
Properties:
Appearance	Clear, yellow liquid
Odor	Citrus
pH (as is)	9.0
Viscosity, cps @ 25C	50
Density (lbs/gal)	7.6

Use Instructions: Use as is or dilute with water.
Performnce: Removes grease, tar, chewing gum, and other oily
    soils effectively.

SOURCE: Stepan Co.: Formulation No. 128

LIMONENE BASED HOUSEHOLD CLEANER

RAW MATERIALS   % By Weight

d-Limonene   30.0
NINOL 11-CM   20.0
Butyl Carbitol   10.0
Na4 EDTA (active)   1.0
Water, soft   39.0

Mixing Procedure:
    Add NINOL 11-CM and Butyl Carbitol to d-Limonene, mix until
clear. Combine water and Na4 EDTA and add slowly under high
agitation. Mix until clear.

Properties:
    Appearance   Clear, yellow liquid
    Odor   Citrus
    pH (as is)   9.5
    Viscosity, cps @ 25C   50
    Density (lbs/gal)   8.0

Use Instructions: Use as is or dilute with water.

Performance:
    Test: GARDNER
    Soil: Oily/particulate
    Substrate: Vinyl tiles
    Concentration of the cleaning solution: 2 oz/gal.
        Product   % Soil removed
    Above formulation   70.0
    Janitor in a drum   32.0
    Top Job   22.0
    Ajax   22.0
    Mr. Clean   20.0

SOURCE: Stepan Co.: Formulation No. 133

LIMONENE BASED HOUSEHOLD AND INDUSTRIAL CLEANER

RAW MATERIALS	% By Weight
d-Limonene	30.0
NINOL 11-CM	20.0
Butyl Carbitol	10.0
Na4 EDTA (active)	2.0
Water, soft	38.0

Mixing Procedure:
   Add NINOL 11-CM and Butyl Carbitol to d-Limonene, mix until
clear. Combine water and Na4 EDTA and add slowly under high agitation. Mix until clear.

Properties:

Appearance	Clear, yellow liquid
Odor	Citrus
pH (as is)	9.5
Viscosity, cps @ 25C	50
Density (lbs/gal)	8.0

Use Instructions: Use as is or dilute with water.

Performance: Remains clear on dilution with water

SOURCE: Stepan Co.: Formulation No. 135

d-LIMONENE SPRAY CLEANER

RAW MATERIALS	% by Weight
Phase A:	
Water, D.I.	68.0
Tetrapotassium Pyrophosphate	6.0
Sodium Metasilicate, Anhydrous	2.0
PETRO LBA Liquid	10.0
DESONATE AOS	3.0
Phase B:	
DESONIC 9N	5.0
Varamide MA-1	3.0
d-Limonene	3.0

Blending Procedure: Blend Phase A and Phase B separately. With
   mixer at high speed, add Phase B to Phase A.

Dilution Ratio: Use as is.

SOURCE: DeSoto, Inc.: Formulation: 3/88:I-3075

LIQUID CLEANSER

RAW MATERIALS	% By Weight
VARION CADG HS	10.0
Monoethanolamine lauryl sulfate (30%)	20.0
VAROX 1770	2.0
Water	68.0

Mixing Procedure:
  Heat the water to 100-120F.
  Add the ingredients to the water slowly in order shown.

SOURCE: Sherex: Industrial Formulation 38:02.2.2

LIQUID GENERAL PURPOSE ALKALINE STEAM CLEANER--HEAVY DUTY

RAW MATERIALS	% By Weight
Water	26.3
Sodium Hydroxide (50%)	6.0
Potassium Hydroxide (45%)	6.7
Reworyl NXS 50	2.0
Sodium Metasilicate	44.0
VARION AMKSF 40	15.0

Mixing Procedure:
  Mix together KOH and NaOH to water. Heat solution and mix
in Sodium Metasilicate. After mixing silicate add Reworyl NXS 50.
Let solution cool and add VARION AMKSF to solution.

SOURCE: Sherex: Industrial Formulation 36:5.5.2

MULTI PURPOSE CLEANER
STRIPPER--CLEANER--DEGREASER

RAW MATERIALS	% By Weight	CAS Registry Number
Water	87.0	7732-11-5
Potassium Hydroxide (90%)	2.4	1310-58-3
Trisodium Phosphate	0.6	7601-54-9
Sodium Metasilicate	3.0	10213-79-3
ESI-TERGE 320	2.0	Mixture
Butyl Cellosolve	5.0	111-76-2
	100.0	

Specifications:
  % Solids   8.0
  % Active   13.0
  pH   13.0-14.0
  Viscosity   Low

SOURCE: Emulsion Systems Inc.: Technical Service Bulletin
        Code HA-20-2

MULTI-PURPOSE CLEANER BASED ON LIMONENE

RAW MATERIALS	% By Weight
d-Limonene	35.0
Deodorized kerosene	53.0
NINATE 411	5.0
MAKON 4	3.0
MAKON 10	2.0
Lanolin	2.0

Mixing Procedure:
   Charge tank with limonene and kerosene. Add remaining ingredients in the order shown above while mixing slowly. Continue mixing until clear.

Properties:
Appearance	Clear liquid
Odor	Citrus
Sp. Gr	0.8

Use Instructions:
   Apply with cloth or sponge on surface to be cleaned, wait a few minutes then wipe clean. Do not allow to dry on.

Performance:
   Removes grease, grime, tar, crayon and lipstick marks, chewing gum and tape adhesive residue from most surfaces without damaging the surface or affecting the paint. Can be used on clothing, carpets, upholstery, bathroom surfaces, appliances, floors, walls and woodwork.

SOURCE: Stepan Co.: Formulation No. 114

GENERAL PURPOSE CLEANER

RAW MATERIALS	% By Weight
SLES (28%)	42.0
VARAMIDE 6CM	5.0
VARAMIDE A-2	4.0
Potassium pyrophosphate	5.0
Water	qs100

Mixing Procedure:
   Dissolve the phosphate into the water followed by the A2 and the 6CM. Finally add the SLES to clear and finish product.

SOURCE: Sherex: Formulation 15:01.6

6. Laundry Products

<u>COLD WATER DETERGENT</u>
(WOOL AND DELICATE FABRICS)

RAW MATERIALS   % By Weight

MIRANOL CS Conc.   10.0
Cedepal SN 303   30.0
Cedemide AX   4.0
Fluorescent Whitening Agent   0.075-0.15
Water   Q.S.

Procedure:
    The fluorescent whitener is solubilized in the MIRANOL CS
CONC. and the Cedepal SN 303 by warming the mixture. The
remaining ingredients are then added.

SOURCE: Miranol Inc.: MIRANOL Products for Household/Industrial
        Applications: Suggested Formulation

<u>COMMERCIAL LAUNDRY LIQUID</u>

RAW MATERIALS   % By Weight

DIACID H-240   6.0
Potassium silicate (1.6:1)   10.0
NaLAS   10.0
Neodol 25-7   5.0
Ethanol   2.0
Water   q.s.*

    *q.s. - quantity sufficient to make 100% total

SOURCE: Westvaco Chemical Division: DIACID Surfactants:
        Suggested Formulation

DETERGENT FOR FINE FABRICS
Clear, Liquid, Without Phosphate

RAW MATERIALS   % By Weight

HOSTAPUR SAS 60   28.0
Lauryletnersulfate-Na (28%)   15.0
Betaine (Tego Betaine L 7)   4.0
Cocofattyacid Diethanolamide   2.2
Potassium Soap (Coco, 27%)   4.5
Water   49.0

Production Procedure:
    Mix HOSTAPUR SAS, LES, Betaine and Cocofattyacid Diethanol-
amide. First add soap and then water, stirring constantly.

Tests:
    pH Value (10% Aqueous Solution T.Q.)   10.0
    Viscosity   390 MPAS
    Stability (-5C)   O.K.
    Freeze and Thaw Test   Clear

SOURCE: Hoechst/Celanese: Formulation A-4004

DETERGENT FOR FINE FABRICS
Liquid, Without Phosphate

RAW MATERIALS   % By Weight

HOSTAPUR SAS 60   22.0
Laurylethersulfate, 28%   5.0
Isotridecylalcoholethoxilate (8EO)   3.0
Coco Fattyacid Diethanolamide   2.0
Perfume Oil   0.3
Preservative Agent, Dye and Water   AD 100%

SOURCE: Hoechst/Celanese: Suggested Formulation

DETERGENT FOR FINE FABRICS
Liquid, With Phosphate

RAW MATERIALS   % By Weight

HOSTAPUR SAS 60   20.0
Laurylethersulfate, 28%   11.0
Nonylphenolethoxilate (10 EO)   2.0
Oleic Acid   2.7
Potassium Hydroxide, 50%   1.2
Sodium Tripolyphosphate   5.0
Urea   2.0
Preservative Agent, Perfume, Dye and Water   AD 100%

SOURCE: Hoechst/Celanese: Suggested Formulation

FINE FABRIC WASH DETERGENT

RAW MATERIALS   % By Weight

Water, d.i.   48.5
Etnanol 3A   7.6
BIO SOFT EA-10   22.7
STEOL CS-460   21.2

Mixing Procedure:
    Add Ethanol 3A to the water. Slowly add BIO SOFT EA-10 to
the mixing solution. Slowly add STEOL CS-460 and mix to a
homogeneous solution. Adjust pH with citric acid.

Properties:
    Appearance   clear, light straw liquid
    pH   7.7
    Viscosity @ 25C, cps   90-120
    Cloud point   <40F
    Freeze/thaw 3 cycles   pass
    130F for 30 days   pass
    Solids, %   34.5

Performance:
    Ross-Miles 0.1% conc: flash foam - 10 cm
                            foam after 5 min - 9 cm
        Draves Wetting Test 0.1% conc: wets in 34 seconds

Comments: A good biodegradable liquid laundry detergent that
          can also be used for baby clothes in the washing
          machine.

SOURCE: Stepan Co.: Formulation No. 116

DETERGENT PASTE WITH PHOSPHATE

RAW MATERIALS   % By Weight

HOSTAPUR SAS 60   43.6
Alkylphenolethoxilate (4EO)   5.0
Cocofattyacid Diethanolamide   0.7
Tylose CBR 4000 (CMC)   5.0
Sodium Tripolyphosphate,   3.0
Sodium Chloride   5.0
Water, Preservative   37.7

Production Procedure:
   Mix Alkylphenolethoxilate, the Cocofattyacid Diethanolamide
and the CMC for about 5 min. Add NACL, preservative and the
phosphate and mix for about 3 min. with low agitation. Let
stand for 24 hrs, then slowly add HOSTAPUR SAS (stirring for
about 10 min). Leave for 15 min., then homogenize again for
10 min.

SOURCE: Hoechst/Celanese: Formulation A-9001

DETERGENT PASTE WITHOUT PHOSPHATE

RAW MATERIALS   % By Weight

HOSTAPUR SAS 30   43,6
Lauryletnersulfate (28%)   10,7
Alkylphenolethoxlat + 4 EO   2,0
Cocofattyacid Diethanolamide   0,7
Tylose CBR 10 000 (CMC)   5,0
Sodium Chloride   8,0
Perfume   0,2
Water, Preservative   29,8

Production Procedure:

   Mix CMC, Nonylphenol, Laurylethersulfate and the Cocofatty-
acid-diethanolamide for some minutes than add water. Mix for
about 10 min., and wait until the next day.
   Than add HOSTAPUR SAS 30 and sodium chloride.

SOURCE: Hoechst/Celanese: Formulation A-9002

DETERGENT-SOFTENER
1/4-Cup Formulation
Formulation A

Active, wt.%: 35.0

RAW MATERIALS	% By Weight
SURFONIC HDL	25.6
Softener: ARMOSOFT WA-104	7.8
Anionic Surfactant: ULTRAWET 45KX	7.8
Optical Brightener	As Desired
Dye and Fragrance	As Desired
Ethanol	8.0
Water	To 100

DETERGENT-SOFTENER
1/4-Cup Formulation
Formulation B

Active, wt. %: ·32.4

RAW MATERIALS	% By Weight
SURFONIC HDL	23.3
Softener: ADOGEN 470--75%	8.0
Triethanolamine	2.3
Optical Brightener: TINOPAL CBS-X	0.35
Dye and Fragrance	As Desired
Ethanol	As Needed
Water	To 100

DETERGENT-SOFTENER
1/4-Cup Formulation
Formulation C

Active, wt. %: 36.0

RAW MATERIALS	% By Weight
SURFONIC N-85	30.0
Softener: VARISOFT 222LT--90%	6.7
Optical Brightener: Tinopal RBS-200%	0.35
Dye and Fragrance	As Desired
Ethanol	As Needed
Water	To 100

SOURCE: Texaco Chemical Co.: Liquid Detergent-Softener Formulations with SURFONIC N Nonionic Surfactants: Table I

DETERGENT-SOFTENER
1/2-Cup Formulation
Formulation D

Active, wt. %: 28.0

RAW MATERIALS   % By Weight

SURFONIC HDL   20.5
Softener: ARMOSOFT WA-104   6.2
Anionic Surfactant: ULTRAWET 45KX   6.2
Optical Brightener   As Desired
Dye and Fragrance   As Desired
Ethanol   9.0
Water   To 100

DETERGENT-SOFTENER
1/2-Cup Formulation
Formulation E

RAW MATERIALS   % By Weight

SURFONIC HDL   18.2
Softener: VARISOFT 3690--75%   4.7
Sodium Citrate Dihydrate   0.45
Optical Brightener: Tinopal CBS-X   0.15
Dye and Fragrance   As Desired
Ethanol   As Needed
Water   To 100

DETERGENT-SOFTENER
1/2-Cup Formulation
Formulation F

Active, wt. %: 21.6

RAW MATERIALS   % By Weight

SURFONIC N-85   18.1
Softener: VARISOFT 222LT-90%   3.9
Sodium Citrate Dihydrate   0.45
Optical Brightener: Tinopal CBS-X   0.15
Dye and Fragrance   As Desired
Ethanol   As Needed
Water   To 100

SOURCE: Texaco Chemical Co.: Liquid Detergent-Softener
        Formulations with SURFONIC N Nonionic Surfactants:
        Table I

DRY-BLENDED LAUNDRY POWDER CONTAINING PHOSPHATE-PREMIUM QUALITY
High Density-One-Quarter Cup

RAW MATERIALS	% By Weight
NEODOL 23-6.5*	13
NEODOL 23-3*	7
Sodium tripolyphosphate	73
Sodium silicate	5
Antiredeposition agent	2
Fluorescent whitening agent**	as desired

Properties:
    Powder density, gm/cc   0.6-0.8

DRY-BLENDED LAUNDRY POWDER CONTAINING PHOSPHATE--HIGH QUALITY
High Density-One-Quarter Cup

RAW MATERIALS	% By Weight
NEODOL 23-6.5*	13
NEODOL 23-3*	7
Sodium tripolyphosphate	58.4
Sodium carbonate***	14.6
Sodium silicate	5
Antiredeposition agent	2
Fluorescent whitening agent**	as desired

Properties:
    Powder density, gm/cc   0.6-0.8

Note:
    If desired, enzymes (e.g., 0.75%w) can be included in these
formulas.

    * The combination of NEODOL 23-6.5 and NEODOL 23-3 may be
      replaced with NEODOL 23-5.
    ** A fluorescent whitening agent should also be included
       (0.1-0.3%w).
    *** Light density soda ash may also be used.

SOURCE: Shell Chemical Co.: The NEODOL Formulary: Suggested
        Formulas

DRY-BLENDED LAUNDRY POWDER CONTAINING PHOSPHATE-PREMIUM QUALITY
High Density-One-Half Cup

RAW MATERIALS	% By Weight
NEODOL 25-7	10
Sodium tripolyphosphate	68
Sodium carbonate*	16.5
Sodium silicate	5
CMC	0.5
Fluorescent whitening agent**	as desired

Properties:
Powder density, gm/cc   0.6-0.8

DRY-BLENDED LAUNDRY POWDER CONTAINING PHOSPHATE--GOOD QUALITY

RAW MATERIALS	% By Weight
NEODOL 25-7	10
Sodium tripolyphosphate	50
Sodium carbonate*	22.5
Sodium silicate	5
Sodium sulfate	12
CMC	0.5
Fluorescent whitening agent**	as desired

Properties:
Powder density, gm/cc   0.6-0.8

Note:
    If desired, enzymes (e.g.,0.75%w) can be included in these formulas.

    * Light density soda ash may also be used.
    ** A fluorescent whitening agent should also be included
       (0.1-0.3%w)

SOURCE: Shell Chemical Co.: The NEODOL Formulary: Suggested
        Formulations

DRY-BLENDED LAUNDRY POWDER CONTAINING PHOSPHATE--GOOD QUALITY
High Density-One-Half Cup

RAW MATERIALS	% By Weight
NEODOL 25-7	9
DDBSA***	3
Sodium tripolyphosphate	50
Sodium carbonate*	11.5
Sodium silicate	5
Sodium sulfate	22
CMC	0.5
Fluorescent whitening agent**	as desired

Properties:
   Powder density, gm/cc   0.6-0.8

DRY-BLENDED LAUNDRY POWDER CONTAINING PHOSPHATE--REGULAR
QUALITY
High Density--One Half Cup

RAW MATERIALS	% By Weight
NEODOL 25-7	10
Sodium tripolyphosphate	34
Sodium carbonate*	23
Sodium silicate	5
Sodium sulfate	27.5
CMC	0.5
Fluorescent whitening agent**	as desired

Properties:
   Powder density, gm/cc   0.6-0.8

Note:
   If desired, enzymes(e.g.,0.75%w) can be included in these
formulas

   * Light density soda ash may also be used
  ** A fluorescent whitening agent should also be included
     (0.1-0.3%).
 *** Dodecylbenzene sulfonic acid

SOURCE: Shell Chemical Co.: The NEODOL Formulary: Suggested
        Formulations

HAND-WASH LAUNDRY LIQUID-MODERATE FOAM

RAW MATERIALS	% By Weight
NEODOL 23-6.5	10.0
NEODOL 25-3S (60%)	12.0
FADEA*	3.0
Sodium chloride	1.0
Water, dye, perfume	to 100%

Properties:
    Viscosity, 73F, cps   451
    Clear point, F   45
    Adjust pH to 7.5-8.0 with citric acid.

   * Fatty acid amide

HAND-WASH LAUNDRY LIQUID--HIGH FOAM, EXTRA MILD

RAW MATERIALS	% By Weight
NEODOL 25-3S (60%)	25.0
FADEA*	3.0
Sodium chloride	2.5
Water, dye, perfume	to 100%

Properties:
    Viscosity, 73F, cps   159
    Clear point, F   46
    Adjust pH to 7.5-8.0 with citric acid.

   * Fatty acid amide

Blending Procedure:
    Dissolve the salt in the water. Add the NEODOL 23-6.5 with mixing. Add the NEODOL 25-3S slowly and with good mixing. Add the amide as a liquid--premelt if necessary.

SOURCE: Shell Chemical Co.: The NEODOL Formulary: Suggested
        Formulations

HEAVY DUTY COMMERCIAL LAUNDRY DETERGENT

RAW MATERIALS   % By Weight

Anhydrous Sodium Metasilicate	87
ESI-TERGE 320	4
Optical Brightener	(.1-.3)
Anhydrous Sodium Tripolyphosphate	9
	100

Procedure:
    To a suitable powder mixer add sodium metasilicate and
ESI-TERGE 320. Allow the mix 10 minutes and add the brightener
and sodium tripolyphosphate. Allow an additional 10 minutes
of mixing.

SOURCE: Emulsion Systems Inc.: Technical Service Bulletin
        Code 320-7

INDUSTRIAL LAUNDRY DETERGENT

RAW MATERIALS   % By Weight

A.

SURFONIC JL-80X	20
CMC	1
Water	79
Cloud point, C, 1% soln.	60

B.

Sodium silicate (2.4:1 $SiO_2:Na_2O$)	5
Potassium hydroxide	14
Tetrapotassium pyrophosphate	10
Water	71

    Parts A and B are mixed separately. Add to the wash water as
1 part A to 4 parts B.

SOURCE: Texaco Chemical Co.: Formulation 1

HEAVY DUTY LIQUID DETERGENT
(TRANSPARENT, WITHOUT BUILDER)

RAW MATERIALS	% By Weight
HOSTAPUR SAS 60	20.0
Nonylphenolethoxilate (8 EO)	20.0
Perfume	0.3
Water, Preservative, Dye	59.7

Production Procedure:
   Mix HOSTAPUR SAS and Nonionic with perfume. Then add water, preservative and dye.

Tests:
pH Value (10% Aqueous Solution T.Q.)	7.1
Viscosity	280 MPAS
Stability (-5C)	Transparent
Freeze and Thaw Test	O.K.

SOURCE: Hoechst/Celanese: Formulation A-200

HEAVY DUTY POWDER

RAW MATERIALS	% By Weight
HOSTAPUR SAS 60	10.0
Tallow Alcohol Ethoxilate (11 EO)	4.0
Tallow Soap	3.0
Sodium Trisilicate	5.0
Sodium Tripolyphosphate	35.0
Carboxymethylcellulose	2.0
Sodium Perborate	25.0
Water	Approx. 10.0
Perfume Oil, Brightener, Dye and Sodium Sulfate	AD 100%

   (All ingredients are calculated as 100% active matter)

SOURCE: Hoechst/Celanese: Suggested Formulation

HEAVY DUTY LIQUID LAUNDRY DETERGENT "HDL"

RAW MATERIALS   % By Weight

Water, d.i.   29.000
STEPANATE X   15.000
Alcohol 3A   4.000
Urea   9.000
PVP-K30   0.100
Tinopal CBS-X   0.100
BIO SOFT LD-190   38.800
Sodium Citrate   4.000
Dye, coloring, preservative as required
TOTAL   100.000

Mixing Procedure:
   Blend ingredients in order above.

Properties:
   Appearance   clear liquid
   pH, as is   8.5-9.5
   Specific Gravity (25C)   1.08-1.10
   Viscosity @ 25C, cps   80-90
   Solids, %   51-54

Use Instructions: 1/2 cup per washload

SOURCE: Stepan Co.: Formulation No. 62

HEAVY DUTY LIQUID WITHOUT PHOSPHATE

RAW MATERIALS   % By Weight

HOSTAPUR SAS 60   15.0
Nonylphenolethoxilate (8EO)   13.0
Potassium Coco Soap, 27%   40.0
Propylenglycol   2.0
Enzymes   0.5
Brightener   0.2
Perfume Oil   0.3
Preservative Agent, Dye and Water   AD 100%

SOURCE: Hoechst/Celanese: Suggested Formulation

HEAVY DUTY LIQUID LAUNDRY DETERGENT(36%) WITH FABRIC SOFTENER(2887-052)

RAW MATERIALS	% By Weight
Water	to 100
Tetrasodium EDTA (40%)	0.5
Ethanol	12.0
Triethanolamine	2.0
**Softener base--VARISOFT 3690 (90%)	6.5
TRYCOL 6964 POE (9) Nonylphenol	28.0
Fragrance, fluorescent whitening agent, dye and preservative	as desired

** Available at 75% concentration with 25% isopropanol.

HEAVY DUTY LIQUID LAUNDRY DETERGENT(36%) WITH FABRIC SOFTENER(2887-052)

RAW MATERIALS	% By Weight
Water	to 100
Tetrasodium EDTA (40%)	0.5
Triethanolamine	2.0
**Softener base--VARISOFT 3690 (90%)	6.5
Sodium xylene sulfonate (40%) (SXS)	10.5
TRYCOL 6964 POE (9) Nonylphenol	25.0
Fragrance, fluorescent whitening agent, dye and preservative	as desired

**Available at 75% concentration with 25% isopropanol.

Blending Procedure:
Charge room temperature water (about 70F) to the blending tank. These formulas should not have to be heated unless heat aids in pumping. While mixing, add the ingredients to the blending tank in the order listed. Stir until uniform. The etnanol or sodium xylene sulfonate is used to modify the viscosity so the product can be pumped. Also, these hydrotropes add stability to the product. A viscosity of less than 300 cP is recommended so that the final product empties from the user's measuring cup. Tinopal CBS-X may be used as a fluorescent whitening agent at approximately 0.35%.

Use Dilution:
Use 1/4 cup per wash load. If a 1/2 cup product is desired, reduce levels of TRYCOL and VARISOFT to 15% and 3%, respectively. Lower the SXS or ethanol to maintain the viscosity required.

SOURCE: Emery Chemicals: Specialty Chemicals Formulary: Formulation 2887-052(3B) and (4B)

HEAVY DUTY LIQUID WITH PHOSPHATE

RAW MATERIALS   % By Weight

HOSTAPUR SAS 60   17.0
Nonylphenolethoxilate (10 EO)   5.5
Potassiumcoco Soap, 40%   10.0
Hydrotrope (HOE S 2817)   5.0
Potassium Hydroxide, 85%   3.0
Potassium Tripolyphosphate, 50%   30.0
Sodium Metasilicate-5 H2O   5.0
Perfume Oil   0.5
Preservative Agent, Brightener, Dye and Water   AD 100%

SOURCE: Hoechst/Celanese: Suggested Formulation

HEAVY DUTY LIQUID DETERGENT
Transparent, Containing Builder(Sodium Citrate)

RAW MATERIALS   % By Weight

HOSTAPUR SAS 60   30.0
Nonylphenolethoxilate (8EO)   12.0
Sodium Citrate   10.0
Hydrotrope (HOE S 2817)   4.0
Perfume   0.3
Water, Preservative, Dye   43.7

Production Procedure:
   Dissolve sodium citrate in water and add a mixture of HOSTAPUR SAS, nonionic and perfume. Then add hydrotrope, preservative and dye, stirring constantly.

Tests:
    pH Value (10% Aqueous Solution T.Q.)   11.3
    Viscosity (Brookfield RVT)   115 MPAS
    Stability (-5C)   Transparent
    Freeze and Thaw Test   O.K.

SOURCE: Hoechst/Celanese: Suggested Formulation A-100

HEAVY-DUTY PHOSPHATE FREE LAUNDRY POWDER-A

RAW MATERIALS	% By Weight
MAZER MACOL 25	12
Soda Ash, Light, Anhydrous	65
Sodium Metasilicate, Liquid	20
Carboxymethyl Cellulose	2
Sodium Sulfate, Brightener, Etc.	1

SOURCE: Mazer Chemicals, Inc.: Household/Industrial T-20B:
Formulation 18A

HEAVY-DUTY PHOSPHATE FREE LAUNDRY POWDER-B

RAW MATERIALS	% By Weight
MAZER MACOL 25	12
Soda Ash, Light, Anhydrous	50
Sodium Metasilicate, Anhydrous Powder	16
Carboxymethyl Cellulose	2
Sodium Sulfate, Brightener, Etc.	20

SOURCE: Mazer Chemicals, Inc.: Household/Industrial T-20B:
Formulation 18B

HEAVY-DUTY POWDERED LAUNDRY DETERGENT

RAW MATERIALS	% By Weight
MAZER MACOL 25	12
Carboxymethyl Cellulose	2
Sodium Tripolyphosphate	34
Sodium Metasilicate, Anhydrous Powder**	8
Sodium Sulfate, Brighteners, Etc.	44

** Na_2O: SiO_2 ratio 1:2--1:2:4,
80-85% solids

SOURCE: Mazer Chemicals, Inc.: Household/Industrial T-20B:
Formulation 16

INDUSTRIAL LAUNDRY DETERGENT

RAW MATERIALS	% By Weight
A.	
SURFONIC N-102	10
SURFONIC N-120	10
CMC	1
Water	79
Cloud Point, C, 1% soln.	76
B.	
Sodium silicate (2.4:1 SiO2:Na2O)	5
Potassium hydroxide	14
Tetrapotassium pyrophosphate	10
Water	71

Parts A and B are mixed separately. Add to the wash water as 1 part A to 4 parts B.

INDUSTRIAL LAUNDRY DETERGENT

RAW MATERIALS	% By Weight
A.	
SURFONIC N-102	20
CMC	1
Water	79
Cloud point, C, 1% soln.	71
B.	
Sodium silicate (2.4:1 SiO2: Na2O)	5
Potassium hydroxide	14
Tetrapotassium pyrophosphate	10
Water	71

INDUSTRIAL LAUNDRY DETERGENT

RAW MATERIALS	% By Weight
A.	
SURFONIC N-85	20
CMC	1
Water	79
Cloud point, C, 1% soln.	45
B.	
Sodium silicate (2.4:1 SiO2: Na2O)	5
Potassium hydroxide	14
Tetrapotassium pyrophosphate	10
Water	71

SOURCE: Texaco Chemical Co.: Suggested Formulations 2,3,4

LAUNDRY DETERGENT-COMMERCIAL: HIGH QUALITY BUILT LAUNDRY
LIQUID

RAW MATERIALS	% By Weight
Part A: Stabilizer System:	
Water	57.7
NEODOL 25-9	0.05
GANTREZ AN-149	0.95
Potassium hydroxide (45%)	4.0
Part B: Detergent System:	
Dye and fluorescent whitening agent(s)	As desired
CMC	0.5
Sodium silicate*	4.0
NEODOL 25-9	10.0
Tetrapotassium pyrophosphate	25.0

Properties:
    Viscosity, 73F, cps   309

SOURCE: Shell Chemical Co.: The NEODOL Formulary: Suggested
        Formulation

LAUNDRY DETERGENT

RAW MATERIALS	% By Weight
Sodium carbonate	47
PLURAFAC A-38 surfactant or	
PLURAFAC D-25 surfactant	12
Sodium metasilicate pentahydrate	24
Sodium hydroxide	15
Carboxymethylcellulose	2

SOURCE: BASF Corp.: Formulation #3850

LAUNDRY DETERGENT

RAW MATERIALS	% By Weight
Sodium tripolyphosphate	40
Sodium carbonate	10
PLURAFAC D-25 surfactant	10
Sodium metasilicate pentahydrate	20
Carboxymethylcellulose	0.5
Sodium sulfate	19.2

SOURCE: BASF Corp.: Formulation #3875

LAUNDRY DETERGENTS-COMMERCIAL--UNBUILT LIQUIDS*

UNBUILT LIQUIDS*--PREMIUM QUALITY

RAW MATERIALS	% By Weight
NEODOL 23-6.5	37.5
C12 LAS (60%)**	20.8
Ethanol	6.0
Triethanolamine	3.0
Potassium chloride	1.0
Water, dye, perfume, fluorescent whitening agent(s)	to 100%

Properties:

Viscosity, 73F, cps	105
Phase coalescence temp., F	130
Clear point, F	32
pH	10

UNBUILT LIQUID--GOOD QUALITY

RAW MATERIALS	% By Weight
NEODOL 91-6	15.0
NEODOL 23-6.5	15.0
C12 LAS (60%)**	16.7
Sodium sesquicarbonate	2.0
Ethanol	1.0
Water, dye, perfume, fluorescent whitening agent(s)	to 100%

Properties:

Viscosity, 73F, cps	135
Phase coalescence temp., F	>176
Clear point, F	32
pH	10

* For use at wash temperatures of 80-140F. Add bleach in
  separate step.
** May use the appropriate amount of dodecylbenzene sulfonic
   acid (DDBSA) with an equivalent amount of sodium hydroxide
   to neutralize it.

Blending Procedure:
  For the preparation of unbuilt, clear-type, HDL formulations,
the order of addition is of importance to minimize viscosity
resistance to mixing and to avoid possible gel formation. Effective stirring should be maintained during addition of all
ingredients, and each ingredient should be in solution before
the next is added.

SOURCE: Shell Chemical Co.: The NEODOL Formulary: Suggested
        Formulations

LAUNDRY LIQUID CONCENTRATE

RAW MATERIALS   % By Weight

NEODOL 25-7	15
NEODOL 91-8	15
C12 LAS (60%)	16.7
Sodium carbonate	0.87
Sodium bicarbonate	0.87
Water	51.56

Properties:
Active matter, %W	40
Viscosity, 73F, cps	240
pH	9.8

Blending Procedure:
Dissolve the sodium carbonate and sodium bicarbonate in the
water. With stirring add the linear alkylbenzene
sodium sulfonate (LAS), then the NEODOL 91-8, then the
NEODOL 25-7.

Note:
By addition of fluorescent whitening agent(s), dye and
perfume, this concentrate can be used as a premium laundry
liquid, or it can be diluted as shown below to make less
concentrated products.

LAUNDRY LIQUID FROM CONCENTRATE-GOOD QUALITY

RAW MATERIALS   % By Weight

Concentrate	87.5
Sodium carbonate	0.34
Sodium bicarbonate	0.34
Water, dye, perfume, fluorescent whitening agent(s)	to 100%

Properties:
Viscosity, 73F, cps	220
Clear point, F	43

SOURCE: Shell Chemical Co.: The NEODOL Formulary: Suggested
        Formulations

LAUNDRY LIQUID FROM CONCENTRATE--STORE BRAND

RAW MATERIALS   % By Weight

Concentrate   75.0
Sodium carbonate   0.44
Sodium bicarbonate   0.44
Water, dye, perfume, fluorescent whitening agent(s)   to 100%

Properties:
   Viscosity, 73F, cps   240
   Clear point, F   43

LAUNDRY LIQUID FROM CONCENTRATE--ECONOMY

RAW MATERIALS   % By Weight

Concentrate   62.5
Sodium carbonate   0.28
Sodium bicarbonate   0.28
Water, dye, perfume, fluorescent whitening agent(s)   to 100%

Properties:
   Viscosity, 73F, cps   250
   Clear point, F   43

LAUNDRY LIQUID FROM CONCENTRATE--GENERIC

RAW MATERIALS   % By Weight

Concentrate   50.0
Sodium carbonate   0.63
Sodium bicarbonate   0.63
Water, dye, perfume, fluorescent whitening agent(s)   to 100%

Properties:
   Viscosity, 73F, cps   180
   Clear point, F   43

SOURCE: Shell Chemical Co.: The NEODOL Formulary: Suggested
      Formulations

LAUNDRY LIQUID WITH ENZYMES--PREMIUM QUALITY

RAW MATERIALS	% By Weight
NEODOL 23-6.5*	30
NEODOL 25-3S (60%)	20
Triethanolamine	1
Enzyme system**	1-2
Stabilizer***	1
Ethanol SD-3A	5
Potassium chloride	4
Fluorescent whitening agent(s)	0.3-0.5
Water, dye, perfume	to 100%

Properties:
    Viscosity, 73F, cps   140
    Clear point, F   39
    Temperature stability, 140F for 1 week   pass
    Freeze-thaw test (3 cycles)   pass

Use Concentration: 1/4 cup

LAUNDRY LIQUID WITH ENZYMES--GOOD QUALITY

RAW MATERIALS	% By Weight
NEODOL 23-6.5*	30
C12 LAS (60%)	10
Triethanolamine	1
Enzyme system**	1-2
Stabilizer***	1
Ethanol SD-3A	5
Potassium chloride	2
Fluorescent whitening agent(s)	0.3-0.5
Water, dye, perfume	to 100%

Properties:
    Viscosity, 73F, cps   145
    Clear point, F   18
    Temperature stability,
        140F for 1 week   pass
    Freeze-thaw test (3 cycles)   pass

Use Concentration: 1/4 cup
    * NEODOL 25-7 or 25-9 may be used in place of NEODOL 23-6.5
   ** Protease and/or amylase enzymes may be used.
  *** Stabilizers currently used are patented short chain carbox-
      ylic acid salts (formates, acetates). Their use may violate
      patent rights.

SOURCE: Shell Chemical Co.: The NEODOL Formulary: Suggested
        Formulations

LAUNDRY LIQUID WITH FABRIC SOFTENER-PREMIUM QUALITY

RAW MATERIALS	% By Weight
NEODOL 25-9	26.0
C12 LAS (60%)	6.7
Armosoft WA104	6.0
Triethanolamine	5.0
Potassium chloride	1-2
Ethanol	9-10
Water, dye, perfume, fluorescent whitening agent(s)	to 100%

Properties:

Viscosity, 73F, cps	110
Clear point, F	41

LAUNDRY LIQUID WITH FABRIC SOFTENER-PREMIUM QUALITY*

RAW MATERIALS	% By Weight
NEODOL 25-9	30.0
Armosoft WA104	6.0
Triethanolamine	5.0
Ethanol	10.0
Water, dye, perfume, fluorescent whitening agent(s)	to 100%

Properties:

Viscosity, 73F, cps	180
Clear point, F	63

* This formula is moderately superior for removing mixed sebum soils, but somewhat less effective against oily soils and for preventing soil redeposition.

LAUNDRY LIQUID WITH FABRIC SOFTENER-GOOD QUALITY

RAW MATERIALS	% By Weight
NEODOL 25-9	21.3
C12 LAS (60%)	4.5
Armosoft WA104	4.0
Triethanolamine	3.0
Potassium chloride	2-3
Ethanol	11
Water, dye, perfume, fluorescent whitening agent(s)	to 100%

SOURCE: Shell Chemical Co.: The NEODOL Formulary: Suggested Formulations

LAUNDRY LIQUID WITH FABRIC SOFTENER AND ENZYMES

RAW MATERIALS	% By Weight
NEODOL 25-9	24
ARMOSOFT WA104	5
Triethanolamine	1
Enzyme system**	1
Stabilizer***	1
Ethanol SD-3A	10
Fluorescent whitening agent(s)	0.3-0.5
Water, dye, perfume	to 100%

Properties:
Viscosity, 73F, cps	80
Clear point, F	62
Temperature stability, 140F for 1 week	pass
Freeze-thaw test (3 cycles)	pass

  ** Protease and/or amylase enzymes may be used.
 *** Stabilizers currently used are patented short chain car-
     boxylic acid salts (formates, acetates). Their use may
     violate patent rights.

Blending Procedure:
   For the preparation of unbuilt, clear-type, HDL formulations,
the order of addition is of importance to minimize viscosity
resistance to mixing and to avoid possible gel formation.
Effective stirring should be maintained during addition of all
ingredients, and each ingredient should be in solution before
the next is added. A blending temperature somewhat above ambient
(e.g., 80-90F) is recommended but not essential.

SOURCE: Shell Chemical Co.: The NEODOL Formulary: Suggested
        Formulation

LAUNDRY PASTE

RAW MATERIALS	% By Weight
HOSTAPUR SAS 60	43.6
Laurylethersulfat, 28%	10.7
Nonylphenolethoxilate (4 EO)	2.0
Coco Fattyacid Diethanolamide	0.7
Carboxymethylcellulose	5.0
Sodium Chloride	8.0
Perfume Oil	0.2
Water, Preservative Agent	AD 100%

SOURCE: Hoechst/Celanese: Suggested Formulation

LIGHT DUTY LIQUID DETERGENT
Clear, Liquid, 40% AM

RAW MATERIALS	% By Weight
HOSTAPUR SAS 60	53.3
Laurylethersulfate-Na (28%)	28.6
Ethanol	3.0
Perfume Oil	0.3
Water, Preservative, Dye	14.8

Production Procedure:
   Mix HOSTAPUR SAS and LES with the perfume. Then add water, ethanol, preservative and dye.

Tests:
   pH Value (10% Aqueous Solution T.Q.)   7.8
   Viscosity   Approx. 350 MPAS
   Stability (+-0C)   Clear
   Clear Point   +13C
   Freeze and Thaw Test   O.K.

SOURCE: Hoecnst/Celanese: Formulation C-1006

CONCENTRATED LIQUID LAUNDRY DETERGENT

RAW MATERIALS	% By Weight
Avanel S-70	50.0
MAZER MACOL 25	25.0
MAZER MACOL 45	10.0
MAZER MACOL 48	7.0
MAZER MACOL 41	8.0
Dye	q.s.
Fragrance	q.s.

SOURCE: Mazer Chemicals, Inc.: Household/Industrial T-20B: 23

LAUNDRY POWDERS--COMMERCIAL

LOW TEMPERATURE POWDER*

PHOSPHATE/CAUSTIC

RAW MATERIALS	% By Weight
NEODOL 25-7	7.5
NEODOL 25-3	2.5
Sodium metasilicate, pentahydrate	32.0
Sodium tripolyphosphate, anhydrous basis	18.0
Sodium hydroxide, granular	26.0
Sodium sulfate	13.0
CMC**	1.0

NON-PHOSPHATE/NON-CAUSTIC

RAW MATERIALS	% By Weight
NEODOL 25-7	6.0
NEODOL 25-3	2.0
Sodium metasilicate, pentahydrate	58.0
Sodium carbonate	33.0
CMC**	1.0

PHOSPHATE/NON-CAUSTIC

RAW MATERIALS	% By Weight
NEODOL 25-7	7.5
NEODOL 25-3	2.5
Sodium metasilicate, pentahydrate	39.0
Sodium tripolyphosphate, anhydrous basis	18.0
Sodium sulfate	9.0
Sodium carbonate	23.0
CMC**	1.0

Blending Procedure:
   Mix solid builders and fillers thoroughly. Add non-ionic slowly while mixing, mix thoroughly. Add CMC and fluorescent whitening agents (as desired).

   * For higher use temperature (e.g., 150F or above) replace
     NEODOL 25-7 and NEODOL 25-3 with NEODOL 25-9
  ** Carboxymethylcellulose.

SOURCE: Snell Chemical Co.: The NEODOL Formulary: Suggested
        Formulation

LIGHT DUTY LIQUID

RAW MATERIALS   % By Weight

HOSTAPUR SAS 60   40.0
Laurylethersulfate, 28%   21.4
Perfume Oil   0.2
Preservative Agent, Dye and Water   AD 100%

SOURCE: Hoechst/Celanese: Suggested Formulation

LIGHT DUTY LIQUID
Transparent, 20% A.M.

RAW MATERIALS   % By Weight

HOSTAPUR SAS 60   26.7
Laurylethersulfate (28%)   14.3
Perfume   0.2
NaCl   1.5
Water, Preservative, Dye   57.3

Production Procedure:
   HOSTAPUR SAS 60 and LES are mixed with perfume. Add water,
preservative and dye. Then adjust the viscosity by adding NaCl.
Tests:
      pH Value (10% Aqueous Solution T.Q.)   7.6
      Viscosity   Approx. 230 MPAS
      Clear Point   + 12.5C
      Stability (+-OC)   Transparent
      Freeze and Thaw Test   O.K.

SOURCE: Hoechst/Celanese: Formulation C-1004

LIGHT DUTY LIQUID DETERGENT
Transparent, 35% A.M.

RAW MATERIALS   % By Weight

HOSTAPUR SAS 60   47.5
Laurylethersulfate-NA (28%)   25.0
Perfume   0.2
Water, Preservative, Dye   27.3

Production Procedure:
   Mix HOSTAPUR SAS 60 and LES with the perfume. Add water,
preservative and dye, stirring constantly.
Tests
      pH Value (10% Aqueous Solution T.Q.)   7.4
      Viscosity   510 MPAS
      Stability (-5C)   Transparent
      Freeze and Thaw Test   O.K.

SOURCE: Hoechst/Celanese: Formulation C-1005-1

LIQUID DETERGENT-SOFTENER
1/2 Cup Formulation-I

RAW MATERIALS	% By Weight
Aliphatic Alcohol Ethoxylate (7-9 mole EO)	15.0
VARISOFT 3690-75%	4.0
Tinopal CBS-X	0.2
Ethanol	3.0
Water	77.8
Dye/Fragrance	(As Desired)

LIQUID DETERGENT-SOFTENER
1/2 Cup Formulation-II

RAW MATERIALS	% By Weight
Aliphatic Alcohol Ethoxylate (7-9 mole EO)	15.0
ADOGEN 470-75%	4.0
Tinopal RBS-200	0.2
Ethanol	3.0
Water	77.8
Dye/Fragrance	(As Desired)

LIQUID DETERGENT-SOFTENER
1/2 Cup Formulation-III

RAW MATERIALS	% By Weight
Alkyl Phenol Ethoxylate (8-9 mole EO)	15.0
VARISOFT 3690N-90%	3.0
Tinopal CBS-X	0.2
Ethanol	3.0
Water	77.8
Dye/Fragrance	(As Desired)

LIQUID DETERGENT-SOFTENER
1/2 Cup Formulation-IV

RAW MATERIALS	% By Weight
Alkyl Phenol Ethoxylate (8-9 mole EO)	18.1
VARISOFT 3690-75%	4.7
Tinopal CBS-X	0.15
Water	77.05
Dye/Fragrance	(As Desired)

SOURCE: Sherex: Formulating Liquid Detergent-Softeners: Formulary

LIQUID DETERGENT-SOFTENER
1/2 Cup Formulation-V

RAW MATERIALS	% By Weight
Alkyl Phenol Ethoxylate (8-9 mole EO)	15.2
VARISOFT 3690-75%	4.7
Sodium Citrate Dihydrate	.045
Triethanolamine, 97%	3.0
Tinopal CBS-X	0.15
Water	76.5
Dye/Fragrance	(As Desired)

LIQUID DETERGENT-SOFTENER
1/2 Cup Formulation-VI

RAW MATERIALS	% By Weight
Alkyl Phenol Ethoxylate (8-9 mole EO)	18.1
VARISOFT 222LT-90%	3.9
Sodium Citrate Dihydrate	.045
Tinopal CBS-X	0.15
Water	77.4
Dye/Fragrance	(As Desired)

LIQUID DETERGENT-SOFTENER
1/2 Cup Formulation-VII

RAW MATERIALS	% By Weight
Alkyl Phenol Ethoxylate (8-9 mole EO)	10.0
VARISOFT 222LT-90%	3.4
Triethanolamine, 97%	2.8
Tinopal CBS-X	0.15
Water	83.65
Dye/Fragrance	(As Desired)

SOURCE: Sherex: Formulating Liquid Detergent-Softeners: Formulary

LIQUID DETERGENT-SOFTENER
1/4 CUP FORMULATION-I

RAW MATERIALS	% By Weight
Aliphatic Alconol Ethoxylate (7-9 mole EO)	30
VARISOFT 3690-75%	8.0
Tinopal CBS-X	0.35
Ethanol	6.0
Water	55.65
Dye/Fragrance	(As desired)

LIQUID DETERGENT-SOFTENER
1/4 CUP FORMULATION-II

RAW MATERIALS	% By Weight
Alkyl Phenol Ethoxylate (8-9 mole EO)	30
Adogen 470-75%	8.0
Tinopal RBS-200	0.35
Ethanol	6.0
Water	55.65
Dye/Fragrance	(As desired)

LIQUID DETERGENT-SOFTENER
1/4 CUP FORMULATION-III

RAW MATERIALS	% By Weight
Alkyl Pnenol Ethoxylate (8-9 mole EO)	30
VARISOFT 3690N-90%	6.7
Tinopal CBS-X	0.35
Ethanol	6.0
Water	56.95
Dye/Fragrance	(As desired)

LIQUID DETERGENT-SOFTENER
1/4 CUP FORMULATION-IV

RAW MATERIALS	% By Weight
Alkyl Phenol Ethoxylate (8-9 mole EO)	20
VARISOFT 222LT-90%	6.7
Triethanolamine	5.6
Tinopal CBS-X	0.35
Water	67.35
Dye/Fragrance	(As desired)

SOURCE: Sherex: Formulating Liquid Detergent-Softeners: Formulary

LIQUID DETERGENT

RAW MATERIALS		% By Weight
SURFONIC N-85		34.3
C12 LAS WITCONATE 1238		5.0
Triethanolamine		5.7
Ethanol		6.0
Active, wt. %	45.0	
Properties:		
Viscosity, cs, 25C	224	
60F	520	

Formulation B

LIQUID DETERGENT

RAW MATERIALS		% By Weight
SURFONIC N-85		40.0
C12 LAS WITCONATE 1238		5.0
Triethanolamine		0
Ethanol		8.0
Active, wt. %	45.0	
Properties:		
Viscosity, cs, 25C	231	
60F	608	

Formulation A

LIQUID DETERGENT

RAW MATERIALS		% By Weight
SURFONIC N-85		25.7
C12 LAS WITCONATE 1238		5.0
Triethanolamine		4.3
Ethanol		6.0

Active, wt. %: 35.0

Properties:	
Viscosity, cs, 25C	224
60F	540

Formulation D

SOURCE: Texaco Chemical Co.: Formulating Liquid Laundry
        Products with SURFONIC N Nonionic Surfactants: Table II

LIQUID DETERGENT

RAW MATERIALS	% By Weight
SURFONIC N-85	30.0
C12 LAS WITCONATE 1238	10.0
Triethanolamine	5.0
Ethanol	6.0

Active, wt. %: 45.0

Properties:
Viscosity, cs, 25C   204
           60F   445

Formulation C

LIQUID DETERGENT

RAW MATERIALS	% By Weight
SURFONIC N-85	21.4
C12 LAS WITCONATE 1238	10.0
Trietnanolamine	3.6
Ethanol	6.0

Active, wt. %:   35.0

Properties:
Viscosity, cs, 25C   171
           60F   402

Formulation E

LIQUID DETERGENT

RAW MATERIALS	% By Weight
SURFONIC N-85	17.1
C12 LAS WITCONATE 1238	5.0
Triethanolamine	2.9
Ethanol	4.0

Active, wt. %   25.0

Properties:
Viscosity, cs, 25C   131
           60F   697

Formulation F

SOURCE: Texaco Chemical Co.: Formulating Liquid Laundry
         Products with SURFONIC N Nonionic Surfactants: Table II

LIQUID DETERGENT

RAW MATERIALS	% By Weight
SURFONIC N-85	12.9
C12 LAS WITCONATE 1238	10.0
Triethanolamine	2.1
Ethanol	4.0

Active, wt. %	25.0

Properties:
Viscosity, cs, 25C	121
60F	743

Formulation G

LIQUID DETERGENT

RAW MATERIALS	% By Weight
SURFONIC N-85	30.0
C13 LAS CONOCO C-650	10.0
Triethanolamine	5.0
Ethanol	7.0

Active, wt. %	45.0

Properties:
Viscosity, cs, 25C	212

Formulation O

LIQUID DETERGENT

RAW MATERIALS	% By Weight
SURFONIC N-85	20.0
C13 LAS CONOCO C-650	10.0
Triethanolamine	5.0
Ethanol	7.0

Active, wt. %	35.0

Properties:
Viscosity, cs, 25C	184

Formulation P

SOURCE: Texaco Chemical Co.: Formulating Liquid Laundry
        Products with SURFONIC N Nonionic Surfactants: Table II

LIQUID DETERGENT

RAW MATERIAL		% By Weight
SURFONIC N-95		15.0
C12 LAS ULTRAWET 45KX		7.0
Triethanolamine		3.0
Ethanol		0
Active, wt. %	25.0	
Properties:		
Viscosity, cs, 25C	140	
60F	269	

Formulation H

LIQUID DETERGENT

RAW MATERIALS		% By Weight
SURFONIC N-100		20.0
C12 LAS CONOCO C-560		6.6
Triethanolamine		3.4
Ethanol		0
Sodium Xylene Sulfonate		2.0
Active, wt. %	30.0	
Properties:		
Viscosity, cs, 25C	161	
60F	259	

Formulation L

LIQUID DETERGENT

RAW MATERIALS		% By Weight
SURFONIC N-102		20.0
C12 LAS CONOCO C-560		6.6
Triethanolamine		3.4
Ethanol		0
Sodium Xylene Sulfonate		2.0
Active, wt. %	30.0	
Properties:		
Viscosity, cs, 25C	138	
60F	262	

Formulation M

SOURCE: Texaco Chemical Co.: Formulating Liquid Laundry
Products with SURFONIC N Nonionic Surfactants: Table II

LIQUID DETERGENT

RAW MATERIALS	% By Weight
SURFONIC N-100	20.0
C12 LAS CONOCO C-560	6.6
Triethanolamine	3.4
Ethanol	0
Sodium Xylene Sulfonate	1.0

Active, wt. %   30.0

Properties:
   Viscosity, cs, 25C   101
           60F   180

Formulation N

LIQUID DETERGENT

RAW MATERIALS	% By Weight
SURFONIC N-85	19.7
C12 LAS ULTRAWET 45KX	10.0
Triethanolamine	3.3
Ethanol	0
Sodium Xylene Sulfonate	4.0

Active, wt. %   33.0

Properties
   Viscosity, cs, 25C   247
           60F   529

Formulation I

LIQUID DETERGENT

RAW MATERIALS	% By Weight
SURFONIC N-85	22.3
C12 LAS ULTRAWET 45KX	10.0
Triethanolamine	3.7
Ethanol	0
Sodium Xylene Sulfonate	5.0

Active, wt. %   36.0

Properties:
   Viscosity, cs, 25C   243
           60F   523

Formulation J

SOURCE: Texaco Chemical Co.: Formulating Liquid Laundry
       Products with SURFONIC N Nonionic Surfactants: Table II

LIQUID DETERGENT

RAW MATERIALS	% By Weight
SURFONIC N-95	24.0
C12 LAS ULTRAWET 45KX	8.0
Triethanolamine	4.0
Ethanol	0
Sodium Xylene Sulfonate	4.0

Active, wt. %	36.0

Properties:

Viscosity, cs, 25C		158
	60F	305

Formulation K

SOURCE: Texaco Chemical Co.: Formulating Liquid Laundry
Products with SURFONIC N Nonionic Surfactants:
Table II

LIQUID LAUNDRY DETERGENT
Heavy Duty Liquid Detergent(33%)(2887-052)

RAW MATERIALS	% By Weight
Water	to 100
Caustic soda (50% sodium hydroxide)	2.3
Dodecylbenzene sulfonic acid (DDBSA)	9.0
Sodium xylene sulfonate (40%) (SXS)	q.s.
	(~10.0)
TRYCOL 6964 POE (9) Nonylphenol	18.0
Tetrasodium EDTA (40%)	0.5
Fluorescent whitening agent	q.s.
Fragrance and preservative	as desired
Triethanolamine or 50% NaOH (to pH 8-10)	q.s.
Opacifier (for Wisk-type product)	as desired
Dye(s)	q.s.

Blending Procedure:
   Charge water to batching tank. While mixing, add raw materials
in the order listed. Warm water will facilitate blending of surfactants. The pH of the batch tank should be higher than 5 after
the DDBSA has been added. If not, adjust with sodium hydroxide
before proceeding.
   Add the SXS to adjust the formula to the desired viscosity.
If an opaque, Wisk-type appearance is desired, an opacifier
(about 0.15%) such as WITCOPAQUE R-25 or E-288 (Morton) may be
used. Use Dilution: Recommended use level: 1/4 cup per washload.

SOURCE: Emery Chemicals: Specialty Chemicals Formulary:
Formulation 2887-052

LAUNDRY DETERGENT, LIQUID-EMULSION TYPE

RAW MATERIALS	% By Weight
Water	42
ACRYSOL ICS-1 Polymer (30%)	2
ACRYSOL ASE-108 Stabilizer (18%)	4
ACRYSOL LMW-45 Polymer (48%)	2
TRITON N-101 Surfactant	10
NaOH (50%)	40

Total Solids (%)	32.28
Appearance	Opaque

LAUNDRY DETERGENT, LIQUID--SLURRY TYPE

RAW MATERIALS	% by weight
Water	32
ACRYSOL ICS-1 Polymer (30%)	2
ACRYSOL ASE-108 Stabilizer (18%)	4
STPP	10
ACRYSOL LMW-45 Polymer (48%)	2
TRITON N-101 Surfactant	10
NaOH (50%)	40

Total Solids (%)	42.28
Appearance	Opaque

Mixing Instructions:
   Add ingredients in the order listed with subsurface agitation.
To prevent air entrapment, avoid mixing at speeds high enough
to form a vortex. Allow sufficient time for hydration of the
STPP before addition of the NaOH.

   Excellent stability and suspendability are provided to these
highly-alkaline formulations by a combination of ACRYSOL ASE-108
and ACRYSOL ICS-1 polymers. Both formulations can be pumped with
a peristaltic pump. ACRYSOL LMW-45 low-molecular-weight polyacrylic acid improves overall cleaning, prevents soil redeposition and, in the slurry formulation, aids in the precipitation
and dispersion of the hydrated STPP. TRITON N-101 surfactant or
blends of TRITON N-101 and TRITON N-60 surfactants provide excellent cleaning performance in these formulations.

SOURCE: Rohm and Haas Co.: Specialty Chemicals: Detergent
        Formulations for Industrial and Institutional Industry:
        Lit. Ref: CS-408/CS-500/CS-504/CS-514

LAUNDRY DETERGENT, LIQUID

RAW MATERIALS	% By Weight
Water	18.9
Carboxymethylcellulose	0.50
Potassium Hydroxide (45% Solution)	2.4
Potassium Silicate (39.4% Solution)	12.70
Water)	6.50
ACRYSOL ASE-108 Stabilizer) Preblend	7.2
Tetrapotassium Pyrophosphate (60% Solution)	41.70
TRITON X-100 Surfactant	10.00
Optical Brightener	0.10
Dye	0.01
Perfume	trace

Properties:

Percent Solids	43
pH	11.7
Bulk Density, lbs./gal.	10.6
Specific Gravity @ 25C	1.27
Viscosity, cps @ 25C	500

Use Dilution:
    1/2 cup for top-loading machine.
    Leaves little residue in measuring cup or reservoir.
    Excellent for automatic injection.

Variations on this formulation have been examined. Lower foaming can be achieved by using TRITON X-114 surfactant. TRITON X-102 or TRITON N-101 surfactants can replace TRITON X-100 surfactant. Formulations having higher levels of foam are obtained by substituting a higher-foaming surfactant for part of the TRITON X-100 surfactant. Optimum foaming is achieved by blending 80 percent TRITON X-100 surfactant with 20 percent lauricdiethanolamide, or 90 percent TRITON X-100 surfactant with 10 percent sodium alkylarylsulfonate. Excessive amounts of anionic surfactants may cause separation.

SOURCE: Rohm and Haas Co.: Specialty Chemicals: Detergent
        Formulations for Industrial and Institutional Industry:
        Lit. Ref.: CS-427, CS-500

LAUNDRY DETERGENT LIQUID, HEAVY DUTY

RAW MATERIALS	% By Weight
TRITON N-101 Surfactant	40.0
Vegetable Potash Soap (19.1%)	52.3
Ethanol (2B)	7.5
Fluorescent Whitening Agent (Tinopal CBS-X)	0.2
Color, Scent	Optional
	100.0

Note:
   Methanol or isopropyl alcohol can be replaced with ethanol.

Use Dilution: 1/4 cup (2 oz.) per load in top loaders.

SOURCE: Rohm and Haas Co.: Specialty Chemicals: Detergent
         Formulations for Industrial and Institutional Industry:
         Lit. Ref: CS-408

LAUNDRY DETERGENT POWDER

RAW MATERIALS	% By Weight
TRITON X-114 Surfactant	8.00
Sodium Tripolyphosphate (STPP)	40.00
Sodium Sulfate	30.00
Soda Ash	16.00
Sodium Silicate (Anhydrous)	5.00
Carboxymethylcellulose	1.00
Optical Brightener	0.05
Perfume	trace
	100.05

Use Dilution:
   1/2 cup per normal washload. Unaffected by hard water.
Either TRITON N-87 or TRITON X-114 are efficient as surfactants
in controlled-foam detergents.

SOURCE: Rohm and Haas Co.: Specialty Chemicals: Detergent
         Formulations for Industrial and Institutional Industry:
         Lit. Ref.: CS-409, CS-443

LAUNDRY DETERGENT POWDER (LOW-FOAM) (A)

RAW MATERIALS   % By Weight

TRITON X-114 Surfactant	5.00
Sodium Tripolyphosphate (STPP)	40.00
Sodium Sulfate	29.80
Borax	14.00
Sodium Silicate (Britesil C-24)	10.00
Carboxymethylcellulose	1.00
Fluorescent Whitening Agent (Tinopal 5BM Extra Conc.)	(0.20)
	100.00

LAUNDRY DETERGENT POWDER (LOW-FOAM) (B)

RAW MATERIALS   % By Weight

TRITON X-114 Surfactant	5.00
Sodium Tripolyphosphate (STPP)	40.00
Sodium Silicate (Britesil C-24)	15.00
Soda Ash	38.80
Carboxymethylcellulose	1.00
Fluorescent Whitening Agent (Tinopal 5BM Extra Conc.)	(0.20)
	100.00

Use Dilution: 1 part in 20 parts water.

Note: Formulation A compares well with commercial products in cost and performance. Borax is an absorbent for TRITON X-114 Surfactant. Formulation B is a less expensive simplification. Either TRITON N-87 Surfactant or TRITON X-114 Surfactant are efficient surfactants in controlled-foam detergents.

SOURCE: Rohm and Haas Co.: Specialty Chemicals: Detergent Formulations for Industrial and Institutional Industry: Lit. Ref.: CS-409, CS-443

LAUNDRY DETERGENT POWDER--LOW PHOSPHATE

RAW MATERIALS	% By Weight
TRITON N-101 Surfactant	10.00
Sodium Carbonate	28.00
Sodium Silicate (Britesil C-20)	12.00
Sodium Tripolyphosphate (STPP)	32.00
Sodium Sulfate	16.62
Carboxymethylcellulose	1.00
Tinopal UNPA	0.30
Tinopal RBS-200%	0.08
	100.00

LAUNDRY DETERGENT POWDER-NO PHOSPHATE

RAW MATERIALS	% By Weight
TRITON N-101 Surfactant	15.00
Sodium Carbonate	50.00
Sodium Silicate (Britesil C-20)	12.00
Sodium Sulfate	21.62
Carboxymethylcellulose	1.00
Tinopal UNPA	0.30
Tinopal RBS-200%	0.08
	100.00

Mixing Instructions:
   Spray TRITON N-101 Surfactant onto sodium carbonate. Add remaining ingredients in listed order. Agitate until free-flowing powder forms.

Properties:
   Appearance   Free-flowing white powder
   pH (At Use Dilution)   10.2

Use Dilution:
   1/4 to 1/2 cup per load. Performance comparable or superior to commercially available home laundry detergents. For lower foam height, replace TRITON N-101 with TRITON N-87 or TRITON X-114. For improved detergency, increase TRITON N-101 content to 20 percent if formulation remains free-flowing.

SOURCE: Rohm and Haas Co.: Specialty Chemicals: Detergent
        Formulations for Industrial and Institutional Industry:
        Lit. Ref.: CS-408, CS-409, CS-443

LAUNDRY PRE-SPOTTER(WATER BASED)

RAW MATERIALS   % By Weight

TRITON N-42 Surfactant   5.5
ACRYSOL A-5 Polyacrylic Acid (25%)   5.0
Deodorized Mineral Spirits   11.0
Iso-octane   22.0
Water   56.5
                                                            100.0

Use as prepared. Shake before using in a trigger spray device.

SOURCE: Rohm and Haas Co.: Specialty Chemicals: Detergent
        Formulations for Industrial and Institutional Industry:
        Lit. Ref: CS-40, CS-506

LAUNDRY PRE-SPOTTER(WATER BASED)

RAW MATERIALS   % By Weight

TRITON N-42 Surfactant   6.5
Trisodium NTA   2.0
Deodorized Mineral Spirits   11.0
Iso-octane   22.0
Water   58.5
                                                            100.0

Use as prepared. Shake before using in a trigger spray device.

SOURCE: Rohm and Haas Co.: Specialty Chemicals: Detergent
        Formulations for Industrial and Institutional Industry:
        Lit. Ref: CS-40

LAUNDRY PRE-SPOTTER(WATER BASED)

RAW MATERIALS   % By Weight

TRITON N-42 Surfactant   5.5
Trisodium NTA   5.0
Deodorized Mineral Spirits   11.0
Iso-octane   22.0
Water   56.5
                                                            100.0

Use as prepared. Shake before using in a trigger spray device.

SOURCE: Rohm and Haas Co.: Specialty Chemicals: Detergent
        Formulations for Industrial and Institutional Industry:
        Lit. Ref.: CS-40

LAUNDRY PRE-SPOTTER(WATER BASED)

RAW MATERIALS   % By Weight

TRITON N-42 Surfactant   5.5
TRITON X-207 Surfactant   2.5
Trisodium NTA   2.5
Deodorized Mineral Spirits   11.0
Iso-octane   22.0
Water   56.5
                                                           100.0

Use as prepared. Shake before using in a trigger spray device.

SOURCE: Rohm and Haas Co.: Specialty Chemicals: Detergent
        Formulations for Industrial and Institutional Industry:
        Lit. Ref.: CS-40/CS-42

LAUNDRY PRESPOTTER

RAW MATERIALS   % By Weight

MAKON 10   4.60
BIO SOFT D-62   3.25
STEPANOL WA-SPECIAL   3.25
Triethanolamine   2.80
Butyl Cellosolve   2.60
Water   83.50
TOTAL   100.00

Mixing Procedure:
    1. Mix water, Butyl Cellosolve, Triethanolamine, and STEPANOL
       WA-SPECIAL.
    2. With agitation, add MAKON 10.
    3. Blend in BIO SOFT D-62. Mix until homogeneous.

Properties:
    Appearance   clear, colorless liquid
    pH as is   10.0
    Viscosity, cps   4

Use Instructions:
    Spray direct from a hand pump-type sprayer

SOURCE: Stepan Co.: Formulation No. 4

LIQUID LAUNDRY DETERGENT
1/4 Cup Use Level; Moderate Sudsing
Unbuilt--Alkyl EO Sulfate Type

RAW MATERIALS   % By Weight

Monoethanolamine   3.5
Fatty alcohol--3EO sulfate   11.5
Ethanol (95%)   9.9
IGEPAL CO-660   23.7
Trisodium citrate   1.1
Water   50.3
                                                                 100.0

Perfume, optical brighteners, colorants, opacifiers or bluing agents added, as desired, replacing water.

Manufacturing Procedure:

1. Dissolve monoethanolamine, fatty alcohol--3EO sulfate, ethanol and IGEPAL CO-660 in approximately two-thirds of the total water.
2. Dissolve trisodium citrate in remaining water prior to addition to main mix.

Physical Properties:
  pH (as is)   12.4
  pH (1%)   10.1
  Viscosity   80 cps
  Specific Gravity   1.02

SOURCE: GAF Corp.: Formulary: Prototype Formulation GAF 5001

LIQUID LAUNDRY DETERGENT
1/4 Cup Use Level; Moderate Sudsing
Unbuilt--Alkylbenzene Sulfonic Acid Type

RAW MATERIALS	% By Weight
Alkylbenzene sulfonic acid	7.8
Sodium hydroxide	1.5
Lauric acid	0.4
Sodium xylene sulfonate	0.7
Triethanolamine	0.3
Ethanol (95%)	5.0
IGEPAL CO-660	33.0
Sodium formate	0.9
Water	50.4
	100.0

Perfume, optical brighteners, colorants, opacifiers or bluing agents added, as desired, replacing water.

Manufacturing Procedure:
1. Dissolve alkylbenzene sulfonic acid in three-fourths of the total amount of water prior to slow neutralization with sodium hydroxide (Note: Sodium hydroxide and sodium xylene sulfonate may be conveniently added from aqueous solution, as supplied by vendors. The method necessitates balancing the total water accordingly.)
2. Add remaining components in the order listed.
3. Dissolve sodium formate in remaining one-fourth of water prior to addition.

Physical Properties:
pH (as is)	7.1
pH (1%)	7.3
Viscosity	200 cps
Specific Gravity	1.05

SOURCE: GAF Corp.: Formulary: Prototype Formulation GAF 5002

LIQUID LAUNDRY DETERGENT
1/4 Cup Use Level; Softening
Anti-Stat Type

RAW MATERIALS	% By Weight
Dimethylditallow ammonium chloride (75% active)	4.9
IGEPAL CO-660	23.9
Water	56.1
Ethanol (absolute)	15.0
POLECTRON 430	0.1
	100.0

Perfume, nonionic optical brighteners and nonionic colorants added, as desired, replacing water.

Manufacturing Procedure:
1. Melt dimethylditallow ammonium chloride in IGEPAL CO-660 at 66C. Cool to 43C and add components in order listed.

Physical Properties:

pH (as is)	4.1
pH (1%)	7.1
Viscosity	80 cps
Specific Gravity	.99

SOURCE: GAF Corp.: Formulary: Prototype Formulation GAF 5004

LIQUID LAUNDRY PRODUCT
1/4 Cup Use Level, Moderate Sudsing
Unbuilt--Alkylaryl EO Sulfate Type

RAW MATERIALS	% By Weight
ALIPAL CO-433	35.7
Ethanol (95%)	10.0
Monoethanolamine	3.5
IGEPAL CO-630	23.0
Trisodium citrate dihydrate	1.0
Water	26.8
	100.0

Perfume, optical brighteners, colorants, opacifiers or bluing agents added, as desired, replacing water.

Manufacturing Procedure:
1. Dissolve ALIPAL CO-433, ethanol and monoethanolamine in two-thirds of the water, then add IGEPAL CO-630.
2. Dissolve trisodium citrate dihydrate in remaining one-third of the total water and blend into main mix.

Physical Properties:

pH (as is)	9.4	pH (1%)	11.1
Viscosity	160 cps	Specific Gravity	1.04

SOURCE: GAF Corp.: Formulary: Prototype Formulation GAF 5003

LIQUID LAUNDRY DETERGENT
1/2 Cup Use Level; Moderate Sudsing
Unbuilt, Pretreater

RAW MATERIALS	% By Weight
Alkylbenzene sulfonic acid	7.9
Sodium hydroxide	1.4
Diethanolamine	5.0
IGEPAL CO-630	23.0
Sodium sulfate	1.5
Water	61.2
	100.0

The amount of sodium sulfate used should be adjusted to allow for free H2SO4 in alkylbenzene sulfonic acid.
    Perfume, optical brighteners, colorants, opacifiers or bluing agents added, as desired, replacing water.

Manufacturing Procedure:
    1. Dissolve alkylbenzene sulfonic acid in all the available water, prior to slow neutralization with sodium hydroxide. (Note: Sodium hydroxide may be conveniently added from aqueous solution, as supplied by vendors. This method necessitates balancing the total water accordingly.)
    2. Add remaining components in the order listed.
Physical Properties:
    pH (as is)   12.9
    pH (1%)   10.5
    Viscosity   300 cps
    Specific Gravity   1.06

SOURCE: GAF Corp.: Formulary: Prototype Formulation GAF 5053

LIQUID LAUNDRY DETERGENT

RAW MATERIALS	% By Weight
Water, deionized	49.0
STEPANATE X	5.0
Ethanol	4.0
Triethanolamine	2.0
BIO SOFT LD-190	40.0
TOTAL	100.0

Properties:
    Appearance   clear yellow liquid
    Viscosity @ 25C, cps   150
    pH, as is   9.0
    Specific Gravity   1.03
    Density, lbs/gal   8.6
Use Information: 1/4 cup per washload

SOURCE: Stepan Co.: Formulation No. 48

LIQUID LAUNDRY DETERGENT

A range of formulations for quarter-cup-dose liquid products follows:

RAW MATERIALS	% By Weight
SURFONIC N-85, N-95, or N-100*	18-36
Triethanolamine*	3-6
Alkylbenzene sulfonate	5-20
Xylene sulfonate	As needed
Ethanol	As needed
Fragrance, dye, brightener	As desired
* Or SURFONIC HDL	21-42

Omit the triethanolamine for neutral products. Detergents based on SURFONIC nonionic surfactants are especially effective against oily soils on synthetic fabrics.
SURFONIC HDL, N-85, and N-95 blend easily with cationic surfactants to give high-active, liquid, detergent-softener-antistatic formulations. Their high detergency and wetting abilities make such formulations effective and economical.

SOURCE: Texaco Chemical Co.: SURFONIC N-Series Surface-Active Agents: Suggested Formulation

HEAVY-DUTY LIQUID LAUNDRY DETERGENT

RAW MATERIALS	% By Weight
MAZER MACOL 25	10.0
Carboxymethyl Cellulose	0.6
Potassium Hydroxide (50% solution)	4.0
Tetrapotassium Pyrophosphate (60% solution)	36.0
Vinyl Ethyl-Maleic Anhydride Copolymer	0.6
Sodium Metasilicate	5.8
Water	43.0

SOURCE: Mazer Chemicals, Inc.: Household/Industrial T-20B: Formulation 17

NON-PHOSPHATE DRY-BLENDED LAUNDRY POWDER--PREMIUM QUALITY
High Density-One-Quarter Cup

RAW MATERIALS	% By Weight
NEODOL 23-6.5*	13
NEODOL 23-3*	7
ZEOLITE 4A	45
Sodium carbonate**	27
Sodium silicate	6
CMC	2
Fluorescent whitening agent***	as desired

Properties:
Powder density, gm/cc   0.6-0.8

NON-PHOSPHATE DRY-BLENDED LAUNDRY POWDER--GOOD QUALITY
High Density-One-Quarter Cup

RAW MATERIALS	% By Weight
NEODOL 23-6.5*	13
NEODOL 23-3*	7
Sodium carbonate**	73
Sodium silicate	5
CMC	2
Fluorescent whitening agent***	as desired

Properties:
Powder density, gm/cc   0.6-0.8

Dry Blending Procedure:
The dry blending procedure that gives the best results in the laboratory with the nonionic surfactant-based high density laundry powders is the following:
   1. Combine all dry components over a 1-2 minute time period while stirring in a Brabender Visco-Corder viscosimeter/paddle mixer.
   2. Heat NEODOL until single-phase liquid; drop-wise, add warm nonionic to dry component mixture, stirring until nonionic is evenly adsorbed onto dry component beads.
Note:
If desired, enzymes (e.g., 0.75%w) can be included in these formulas.

   * The combination of NEODOL 23-6.5 and NEODOL 23-3 may be replaced with NEODOL 23-5
   ** Light density soda ash may also be used
   *** A fluorescent whitening agent should also be included (0.1-0.3%w)

SOURCE: Shell Chemical Co.: The NEODOL Formulary: Formulas

NON-PHOSPHATE DRY-BLENDED LAUNDRY POWDER-PREMIUM QUALITY
High Density-One-Half Cup

RAW MATERIALS	% By Weight
NEODOL 25-7	10
ZEOLITE 4A	34
Sodium carbonate*	34
Sodium silicate	3
Sodium sulfate	18
CMC	1
Fluorescent whitening agent**	as desired

Properties:
Powder density, gm/cc   0.6-0.8

NON-PHOSPHATE DRY-BLENDED LAUNDRY POWDER-GOOD QUALITY
High Density--One-Half Cup

RAW MATERIALS	% By Weight
NEODOL 25-7	10
Sodium carbonate*	68
Sodium silicate	3
Sodium sulfate	18
CMC	1
Fluorescent whitening agent**	as desired

Properties:
Powder density, gm/cc   0.6-0.8

NON-PHOSPHATE DRY BLENDED LAUNDRY POWDER--ECONOMY
High Density-One-Half Cup

RAW MATERIALS	% By Weight
NEODOL 25-7	10
Sodium carbonate*	46
Sodium silicate	3
Sodium sulfate	40
CMC	1
Fluorescent whitening agent**	as desired

Properties:
Powder density, gm/cc   0.6-0.8

Note: If desired, enzymes (e.g., 0.75%w) can be included in these
       formulas.
   * Light density soda ash may also be used
   ** A fluorescent whitening agent should also be included
      (0.1-0.3%w).

SOURCE: Shell Chemical Co.: The NEODOL Formulary: Formulas

POWDERED LAUNDRY DETERGENT
Phosphate Type

RAW MATERIALS   % By Weight

IGEPAL CO-630   10.0
Sodium carbonate (lt. density)   28.0
Sodium tripolyphosphate (lt. density)   32.0
Sodium metasilicate, anhydrous   12.0
Sodium silicate, anhydrous   17.0
Carboxymethylcellulose   1.0
                                                          100.0

Manufacturing Procedure:
1. Mix IGEPAL CO-630 with sodium carbonate until a uniform powder is obtained.
2. Add sodium tripolyphosphate.
3. Mix sodium silicate, anhydrous; sodium metasilicate, anhydrous and carboxymethylcellulose together. Add to main batch.

Physical Properties:
pH (1%)   11.4
Specific Gravity   .82

SOURCE: GAF Corp.: Formulary: Prototype Formulation GAF 5075

POWDERED LAUNDRY DETERGENT
Nonphosphate Type

RAW MATERIALS   % By Weight

GANTREZ AN-119   1.5
Water   9.5
Sodium carbonate (lt. density)   50.0
Sodium sulfate, anhydrous   22.0
Sodium metasilicate, 5H2O   8.5
Carboxymethylcellulose   0.5
IGEPAL CO-630   8.0
                                                          100.0

Manufacturing Procedure:
1. Premix GANTREZ AN-119 in water (80-85C) until clear.
2. Slowly add GANTREZ/water mixture to sodium carbonate, using agitation to avoid lumping.
3. Add remaining components in order. Add IGEPAL CO-630 to avoid lumping. Mix until a homogeneous powder is obtained.

Physical Properties:
pH (1%)   11.3
Specific Gravity   .87

SOURCE: GAF Corp.: Formulary: Prototype Formulation GAF 5077

REGULAR QUALITY BUILT LAUNDRY LIQUIDS*

HIGHLY ALKALINE LIQUID WITH PHOSPHATE

RAW MATERIALS   % By Weight

NEODOL 25-9   3.0
Sodium metasilicate, pentahydrate   7.6
Tetrapotassium pyrophosphate   4.6
Potassium hydroxide (45%)   13.7
CMC**   1.0
TRITON H-66   5.0
Water, dye, perfume, fluorescent whitening agent(s)   to 100%

Properties:
    Viscosity, 73F, cps   23
    Phase coalescence temp., F   172
    Clear point, F   45

HIGHLY ALKALINE LIQUID--NON-PHOSPHATE

RAW MATERIALS   % By Weight

NEODOL 25-9   3.0
Sodium metasilicate, pentahydrate   11.9
Potassium hydroxide (45%)   6.1
Potassium carbonate   3.4
CMC**   1.0
TRITON H-66   5.0
Water, dye, perfume, fluorescent whitening agent(s)   to 100%

Properties:
    Viscosity, 73F, cps   25
    Phase coalescence temp., F   >176
    Clear point, F   48

  * Use where equipment is not capable of metering in the
    wash ingredients separately. Use two-step product with
    metering capable machines.
 ** Carboxymethylcellulose

SOURCE: Shell Chemical Co.: The NEODOL Formulary: Suggested
        Formulations

REGULAR QUALITY BUILT LAUNDRY LIQUIDS*
LOWER ALKALINITY LIQUID WITH PHOSPHATE

RAW MATERIALS   % By Weight

NEODOL 25-9	3.0
Sodium metasilicate, pentahydrate	8.8
Tetrapotassium pyrophosphate	4.0
Potassium carbonate	5.2
CMC**	1.0
TRITON H-66	5.0
Water, dye, perfume, fluorescent whitening agent(s)	to 100%

Properties:
Viscosity, 73F, cps	23
Phase coalescence temp., F	176
Clear point, F	50

LOWER ALKALINITY LIQUID--NON-PHOSPHATE

RAW MATERIALS   % By Weight

NEODOL 25-9	3.0
Sodium metasilicate, pentahydrate	11.5
Potassium carbonate	6.5
CMC**	1.0
TRITON H-66	5.0
Water, dye, perfume, fluorescent whitening agent(s)	to 100%

Properties:
Viscosity, 73F, cps	23
Phase coalescence temp., F	>176
Clear point, F	32

* Use where equipment is not capable of metering in the wash
   ingredients separately. Use two-step product with metering-
   capable machines.
** Carboxymethylcellulose.

SOURCE: Shell Chemical Co.: The NEODOL Formulary: Suggested
        Formulations

TWO-STEP BUILT LAUNDRY LIQUID FORMULATIONS*

Part A. Surfactant Solution
For Low or Regular Temperature Operation

RAW MATERIALS	% By Weight
NEODOL 23-6.5	20.0
Isopropyl alcohol	7.0
CMC**	1.0
Water	72.0

Properties:
Viscosity, 73F, cps	156
Cloud point, 1% soln., F	110
Clear point, F	30

For High Temperature Operation:

NEODOL 25-9	20.0
CMC**	1.0
Water	79.0

Properties:
Viscosity, 73F, cps	54
Cloud point, 1% soln., F	160
Clear point, F	41

   * For automatic dispensing equipment capable of dispensing
     surfactant and builder solutions separately.
  ** Carboxymethylcellulose.

Part B: Builder Solution:

Phosphate/Caustic:	
Sodium metasilicate, pentahydrate	16.8
Tetrapotassium pyrophosphate	9.5
Potassium hydroxide (45%)	30.4
Water	43.3
Phosphate/Non-Caustic:	
Sodium metasilicate, pentahydrate	19.5
Tetrapotassium pyrophosphate	9.0
Potassium carbonate	11.5
Water	60.0
Non-Phosphate/Caustic:	
Sodium metasilicate, pentahydrate	26.4
Potassium hydroxide (45%)	13.5
Potassium carbonate	7.6
Water	52.5
Non-Phosphate/Non-Caustic:	
Sodium metasilicate, pentahydrate	25.5
Potassium carbonate	14.5
Water	60.0

SOURCE: Shell Chemical Co.: The NEODOL Formulary: Formulations

UNBUILT LAUNDRY LIQUID--SUPER PREMIUM QUALITY

RAW MATERIALS	% By Weight
NEODOL 25-7	37.5
C12 LAS (60%)	20.8
Triethanolamine	3.0
Ethanol SD-3A	6.0
Potassium chloride	1.0
Fluroescent whitening agent(s)	0.3-0.5
Water, dye, perfume	to 100%

Properties:
Active matter, %w	50
Viscosity, 73F, cps	135
Clear point, F	54
Pour point, F	5
pH	10.5

UNBUILT LAUNDRY LIQUID--PREMIUM QUALITY

RAW MATERIALS	% By Weight
NEODOL 25-7	32.0
NEODOL 25-3S (60%)	13.3
Triethanolamine	3.0
Etnanol SD-3A	5.0
Potassium chloride	4.0
Fluorescent whitening agent(s)	0.3-0.5
Water, dye, perfume	to 100%

Properties:
Active matter, %w	40
Viscosity, 73F, cps	185
Clear point, F	45
Pour point, F	25
pH	9.2

Blending Procedure:
For the preparation of unbuilt, clear-type, HDL formulations, the order of addition is of importance to minimize viscosity resistance to mixing and to avoid possible gel formation. Effective stirring should be maintained during addition of all ingredients, and each ingredient should be in solution before the next is added. A blending temperature somewhat above ambient (e.g., 80-90F) is recommended but not essential.

SOURCE: Shell Chemical Co.: The NEODOL Formulary: Suggested
        Formulations

UNBUILT LAUNDRY LIQUID--PREMIUM QUALITY

RAW MATERIALS   % By Weight

NEODOL 25-7	30.0
C12 LAS (60%)	16.7
Triethanolamine	3.0
Ethanol SD-3A	5.5
Potassium chloride	1.0
Fluorescent whitening agent(s)	0.3-0.5
Water, dye, perfume	to 100%

Properties:
Active matter, %	40
Viscosity, 73F, cps	175
Clear point, F	27
Pour point, F	13
pH	10.4

UNBUILT LAUNDRY LIQUID--REGULAR QUALITY

RAW MATERIALS   % By Weight

NEODOL 25-7	22.5
C12 LAS (60%)	12.5
Triethanolamine	3.0
Ethanol SD-3A	5.0
Potassium chloride	2.0
Fluorescent whitening agent(s)	0.3-0.4
Water, dye, perfume	to 100%

Properties:
Active matter, %w	30
Viscosity, 73F, cps	140
Clear point, F	48
Pour point, F	21
pH	10.5

Blending Procedure:
    For the preparation of unbuilt, clear-type, HDL formulations, the order of addition is of importance to minimize viscosity resistance to mixing and to avoid possible gel formation. Effective stirring should be maintained during addition of all ingredients, and each ingredient should be in solution before the next is added. A blending temperature somewhat above ambient (e.g., 80-90F) is recommended but not essential.

SOURCE: Shell Chemical Co.: The NEODOL Formulary: Suggested
        Formulation

ONE-HALF CUP BUILT LAUNDRY LIQUID*-HIGH QUALITY**

RAW MATERIALS   % By Weight

NEODOL 25-9***	6.0
NEODOL 25-3S (60%)	30.0
Coconut fatty acid	2.0
Monoethanolamine	2.0
Citric acid, anhydrous	8.0
Sodium hydroxide (50%)	8.2
Sodium xylene sulfonate (40%)	5.0
Water, dye, perfume, fluorescent whitening agent(s)	to 100%

Properties:
Viscosity, 73F, cps	116
Clear point, F	34
Temperature stability	
140F for 1 week	pass
Freeze-thaw test (3 cycles)	pass
pH	8.6

ONE-HALF CUP BUILT LAUNDRY LIQUID*-HIGH QUALITY

RAW MATERIALS   % By Weight

NEODOL 25-9***	6.0
NEODOL 25-3S (60%)	20.0
C12 LAS (60%)	10.0
Coconut fatty acid	2.0
Monoethanolamine	2.0
Citric acid, anhydrous	8.0
Sodium hydroxide (50%)	8.2
Water, dye, perfume, fluorescent whitening agent(s)	to 100%

Properties:
Viscosity, 73F, cps	160
Clear point, F	32
Temperature stability	
140F for 1 week	pass
Freeze-thaw test (3 cycles)	pass
pH	8.6

* If desired, enzymes (e.g., 0.1%w) can be included in these formulas. Protease and/or amylase enzymes.
** NEODOL 25-3S provides better enzyme stability than C12 LAS.
*** May substitute with NEODOL 25-7 or NEODOL 23-6.5.

SOURCE: Shell Chemical Co.: The NEODOL Formulary: Suggested Formulations

ONE-HALF CUP BUILT LAUNDRY LIQUID*-HIGH QUALITY, HIGH FOAM

RAW MATERIALS	% By Weight
NEODOL 25-9***	6.0
NEODOL 25-3S (60%)	20.0
C12 LAS (60%)	10.0
Monoethanolamine	2.0
Citric acid, anhydrous	8.0
Sodium hydroxide (50%)	7.0
Water, dye, perfume, fluorescent whitening agent(s)	to 100%

Properties:
Viscosity, 73F, cps	145
Clear point, F	30
Temperature stability	
140F for 1 week	pass
Freeze-thaw test (3 cycles)	pass
pH	8.8

Blending Procedure:
For the preparation of clear-type, HDL formulations, the order of addition is of importance to minimize viscosity resistance to mixing and to avoid possible gel formation. Effective stirring should be maintained during addition of all ingredients, and each ingredient should be in solution before the next is added. A blending temperature somewhat above ambient (e.g., 80-90F) is recommended but not essential.

   * If desired, enzymes (e.g., 0.1%w) can be included in these formulas. Protease and/or amylase enzymes can be used.
  ** NEODOL 25-3S provides better enzyme stability than C12 LAS.
 *** May substitute with NEODOL 25-7 or NEODOL 23-6.5.

SOURCE: Shell Chemical Co.: The NEODOL Formulary: Suggested
       Formulations

LAUNDRY PRESPOTTER(2878-107)

RAW MATERIALS   % By Weight

TRYCOL 5966 CE Ethoxylated Alcohol   10.0
TRYCOL 5943 POE (12) Tridecyl Alcohol   10.0
Shell Sol 71   15.0
Isopropyl alcohol (IPA)   15.0
Triethanolamine (TEA)   2.0
EMERSOL 211 Oleic Acid   3.0
Dye and fragrance   as desired
Water   to 100

Blending Procedure:
  Add the water to the blending tank. While mixing, add the
ingredients to the blending tank in the order listed. Stir
until uniform. The finished product may be packaged in an
aerosol or pump container.

SOURCE: Emery Chemicals: Specialty Chemicals Formulary:
        Formula 2878-107

LAUNDRY PRE-SPOTTER-AEROSOL

RAW MATERIALS   % By Weight

AEROTHENE TT Solvent   20.0
DOWANOL DPM glycol ether   6.0
isopropanol   10.0
Polyoxyethylene Glyceride Ester   22.0
deodorized kerosene   22.0
propellant A-70   20.0

Suggested Valve: Precision 0.013" stem/0.013" body
Suggested Actuator: Precision 0.016" MBRT

  The mutual solvency of DOWANOL DPM in kerosene, water and
soap allow the stains to be washed out.

SOURCE: Dow Chemical Co.: The Glycol Ethers Handbook:
        Suggested Formulation

AEROSOL-TYPE PRESPOTTER--SOLVENT-BASED--PREMIUM QUALITY

RAW MATERIALS	% By Weight
NEODOL 25-7	15.0
NEODOL 25-3	15.0
SHELL SOL 71 or 72	64.0
Ethanol	4.0
Water*	2.0

AEROSOL-TYPE PRESPOTTER--SOLVENT-BASED--HIGH QUALITY**

RAW MATERIALS	% By Weight
NEODOL 45-2.25	10.0
SHELL SOL 71 or 72	90.0

AEROSOL-TYPE PRESPOTTER--SOLVENT-BASED--HIGH QUALITY***

RAW MATERIALS	% By Weight
NEODOL 25-7	5.0
NEODOL 25-3	5.0
SHELL SOL 71 or 72	89.0
Water* ****	1.0

Blending Procedure:
   Add water last, mix vigorously.

   * To avoid corrosion, a lined can may be needed.
  ** Designed for heavy greasy soils.
 *** For general stains.
**** For an anhydrous product, use 85% SHELL SOL 71 or 72 and
      5% ethanol.

SOURCE: Shell Chemical Co.: The NEODOL Formulary: Suggested
        Formulation

LAUNDRY PRESPOTTER--PUMP SPRAY-TYPE--PREMIUM QUALITY

RAW MATERIALS	% By Weight
NEODOL 25-3*	10.0
NEODOL 23-6.5	10.0
SHELL SOL 71 or 72	20.0
Isopropyl alcohol	12.0
Triethanolamine oleate**	3.5
Water, dye, perfume	to 100%

Properties:
Viscosity, 73F, cps	37
Clear point, F	18

LAUNDRY PRESPOTTER--PUMP SPRAY-TYPE--PREMIUM QUALITY

RAW MATERIALS	% By Weight
NEODOL 25-3*	22.0
SHELL SOL 71 or 72	18.0
Sodium xylene sulfonate (40%)	30.0
Triethanolamine oleate**	2.5
Water, dye, perfume	to 100%

Properties:
Viscosity, 73F, cps	218
Clear point, F	37

  * May substitute with NEODOL 23-3
 ** Can be prepared in situ from triethanolamine and oleic
    acid.

SOURCE: Shell Chemical Co.: The NEODOL Formulary: Suggested
        Formulations

PUMP SPRAY PRESPOTTER--WATER-BASED--GOOD QUALITY

RAW MATERIALS	% By Weight
NEODOL 25-7*	9.75
NEODOL 25-3*	5.25
Sodium xylene sulfonate (40%)	15.0
EDTA salt	0.5
Water, dye, perfume	to 100%

Properties:
Viscosity, 73F, cps	18
Clear point, F	59

PUMP SPRAY PRESPOTTER--WATER-BASED--REGULAR QUALITY

RAW MATERIALS	% By Weight
NEODOL 25-7*	6.5
NEODOL 25-3*	3.5
Sodium xylene sulfonate (40%)	12.5
EDTA salt	0.5
Water, dye, perfume	to 100%

Properties:
Viscosity, 73F, cps	11
Clear point, F	62

PUMP SPRAY PRESPOTTER--WATER-BASED--ECONOMY

RAW MATERIALS	% By Weight
NEODOL 25-7*	5.2
NEODOL 25-3*	2.8
Sodium xylene sulfonate (40%)	8.75
EDTA salt	0.5
Water, dye, perfume	to 100%

Properties:
Viscosity, 73F, cps	19
Clear point, F	62

* The combination of NEODOL 25-7 and NEODOL 25-3 may be replaced with NEODOL 23-5, adjust SXS as needed for stability.

SOURCE: Shell Chemical Co.: The NEODOL Formulary: Suggested Formulations

7. Metal Cleaners

METAL CLEANER, ACID
(Aluminum and Stainless Steel Cleaner)

RAW MATERIALS	% By Weight
Phosphoric Acid (85%)	3.0
Citric Acid	4.0
TRITON X-100 Surfactant	2.0
Methyl Ethyl Ketone	3.0
Water	88.0
	100.0

Mixing Instructions:
   Add acids slowly to water. Add ketone and TRITON X-100
Surfactant. If cold water is used, premix TRITON X-100 Surfactant
with 3 parts warm water.
Use Dilution: 1 to 2 oz./gallon water.

METAL CLEANER, ACID

RAW MATERIALS	% By Weight
Phosphoric Acid (85%)	35.0
Glycolic Acid	1.0
TRITON X-100 Surfactant	1.5
Water	62.5
	100.0

Mixing Instructions:
   Add phosphoric acid and glycolic acid to water, then TRITON
X-100 Surfactant. If cold water is used, premix TRITON X-100
Surfactant with 3 parts warm water.
Use Dilution: 1 to 2 oz./gal. water.

METAL CLEANER, ACID

RAW MATERIALS	% By Weight
Phosphoric Acid (85%)	35.0
Glycolic Acid	1.0
TRITON X-100 Surfactant	1.5
Water	62.5
	100.0

Mixing Instructions:
   Add phosphoric acid and glycolic acid to water, then TRITON
X-100. If cold water is used, premix TRITON X-100 with 3 parts
warm water.
Use Dilution: 1 to 2 oz./gal. water.

SOURCE: Rohm and Haas Co.: Specialty Chemicals: Detergent
        Formulations for Industrial and Institutional Industry:
        Lit. Ref. CS-427

METAL CLEANER-BRIGHTENER FOR ALUMINUM

RAW MATERIALS	% By Weight
Phosphoric Acid (85%)	47.2
TRITON X-100 Surfactant	2.0
Dipropylene Glycol Methyl Ether (DOWANOL DPM)	16.0
Water	34.8
	100.0

Mixing Instructions:
   Slowly add phosphoric acid to water, then remaining ingredients with agitation. If cold water is used, dilute TRITON X-100 with 3 parts warm water before mixing.

Appearance: Clear, almost colorless liquid

Use Dilution: 1 part/20 parts water

METAL CLEANER-BRIGHTENER FOR ALUMINUM

RAW MATERIALS	% By Weight
Phosphoric Acid (85%)	45.0
TRITON X-102 Surfactant	12.0
Dipropylene Glycol Methyl Ether (DOWANOL DPM)	25.0
o-Dichlorobenzene	5.0
Water	13.0
	100.0

Mixing Instructions:
   Slowly add phosphoric acid to water, then add TRITON X-102 Surfactant and DOWANOL DPM, and finally o-dichlorobenzene.

Appearance: Clear, essentially colorless, viscous liquid.

Directions for Use:
   Dilute with 3 parts water. Effcetive for removing varnish and oxides from aluminum and other metal surfaces.

SOURCE: Rohm and Haas Co.: Specialty Chemicals: Detergent
        Formulations for Industrial and Institutional Industry:
        Lit. Ref.: CS-427/CS-407

METAL CLEANER-BRIGHTENER FOR ALUMINUM

RAW MATERIALS	% By Weight
Phosphoric Acid (85%)	47.2
TRITON X-100 Surfactant	2.0
Dipropylene Glycol Methyl Ether (Dowanol DPM)	16.0
Water	34.8
	100.0

Mixing Instructions:
    Slowly add phosphoric acid to water, then remaining ingred-
ients with agitation. If cold water is used, dilute TRITON X-100
with 3 parts warm water before mixing.

Appearance: Clear, almost colorless liquid.

Use Dilution: 1 part/20 parts water.

Lit. Ref.: CS-427

METAL CLEANER-BRIGHTENER (LIGHT DUTY)
FOR ALUMINUM

RAW MATERIALS	% By Weight
TRITON BG-10 Surfactant (70%)	5.0
Sodium Alkyl Naphthalene Sulfonate (Petro AA)	3.0
Sodium Metasilicate (Anhydrous)	3.0
Tetrapotassium Pyrophosphate (TKPP)	3.0
Dipropylene Glycol Methyl Ether (Dowanol DPM)	5.0
Water	81.0
	100.0

Use Dilution: 1 to 2 oz./gallon water

Lit. Ref.: CS-449

SOURCE: Rohm and Haas Co.: Specialty Chemicals: Detergent
        Formulations for Industrial and Institutional Industry

METAL CLEANER-BRIGHTENER
For Aluminum

RAW MATERIALS	% By Weight
Phosphoric Acid (85%)	47.2
TRITON X-100 Surfactant	2.0
Dipropylene Glycol Methyl Ether (DOWANOL DPM)	16.0
Water	34.8
	100.0

Mixing Instructions:
   Slowly add phosphoric acid to water, then remaining ingredients with agitation. If cold water is used, dilute TRITON X-100 with 3 parts warm water before mixing.

Appearance: Clear, almost colorless liquid.

Use Dilution: 1 part/20 parts water

Lit. Ref.: CS-427

METAL CLEANER, HEAVY DUTY

RAW MATERIALS	% By Weight
Water	80.0
Sodium Hydroxide	0.6
TRITON X-102 Surfactant	2.0
Alkylarylsulfonic Acid (98%)	3.0
Tripotassium Phospnate (TKP)	2.2
Sodium Nitrite	0.2
TAMOL SN Dispersant	2.0
Dipropylene Glycol Methyl Ether (DOWANOL DPM)	10.0
	100.0

Mixing Instructions:
   Combine and agitate all ingredients except DOWANOL DPM. When the ingredients are thoroughly dispersed, add the solvent and mix.

Appearance: Clear, water-soluble liquid.

Use Dilution: 1 to 2 oz./gal. water.

Lit. Ref.: CS-71, CS-407

SOURCE: Rohm and Haas Co.: Specialty Chemicals: Detergent Formulations for Industrial and Institutional Industry

METAL CLEANER, HEAVY DUTY

RAW MATERIALS   % By Weight

TRITON RW-100 Surfactant   5.0
Soda Ash   30.0
Sodium Metasilicate (anhydrous)   35.0
Sodium Hydroxide flakes   30.0
                                                          100.0

Mixing Procedure: Preblend surfactant on soda ash, then add
    other builders.

Use Dilution: 2 to 4 oz./gallon water.

Lit. Ref.: CS-450

METAL CLEANER, HEAVY DUTY

RAW MATERIALS   % By Weight

TRITON X-100 Surfactant   2.0
TRITON X-55 Surfactant (50%)   6.0
Potassium hydroxide   12.0
Sodium Metasilicate (Anhydrous)   12.0
Tetrapotassium pyrophosphate (TKPP)   12.0
Water   56.0
                                                          100.0

Use Dilution: Heavy Soils--3 oz./gal. water
              Light Soils--12 oz./gal. water

Lit. Ref.: CS-427, CS-433

METAL CLEANER, HEAVY DUTY LOW-FOAM

RAW MATERIALS   % By Weight

TRITON CF-54 Surfactant   5.0
Sodium Hydroxide   32.0
Sodium Metasilicate (Anhydrous)   32.0
Soda Ash   31.0
                                                          100.0

Use Dilution: 2 to 4 oz./gallon water.

Note: TRITON CF-10, CF-76, or DF-18 can be used as substitutes.

Lit. Ref.: CS-405, CS-413, CS-436

SOURCE: Rohm and Haas Co.: Specialty Chemicals: Detergent
        Formulations For Industrial and Institutional Industry

METAL CLEANER, HEAVY DUTY LOW-FOAM
LOW TEMPERATURE

RAW MATERIALS	% By Weight
TRITON DF-12 Surfactant	5.0
Sodium Hydroxide	15.0
Sodium metasilicate, anhydrous	35.0
Soda Ash	25.0
Tetrasodium pyrophosphate	20.0
	100.0

Use Dilution: 2 to 4 oz./gallon water
    The low cloud point of TRITON DF-12 gives low-foam character-
istics and detergent effectiveness at low temperature (35C or
less). Heating and energy costs become substantially lower.

Lit. Ref.: CS-415

METAL CLEANER, LIGHT DUTY

RAW MATERIALS	% By Weight
TRITON X-100 Surfactant	5.0
TRITON H-66 Surfactant (50%)	4.0
Sodium Metasilicate (Anhydrous)	3.0
Tetrapotassium Pyrophosphate (TKPP)	3.0
Dipropylene Glycol Methyl Ether (DOWANOL DPM)	5.0
TAMOL SN Dispersant	1.0
Water	79.0
	100.0

Use Dilution: 1 to 2 oz./gal. water.

Lit. Ref.: CS-427, CS-433, CS-71

METAL CLEANER, LIGHT DUTY FOR OIL RIGS

RAW MATERIALS	% By Weight
TRITON N-101 Surfactant	7.8
Ninol 1285	4.7
Sodium Tetraborate (Borax)	2.6
Tetrasodium Ethylenediaminetetraacetate (Versene 100)	1.6
Dipropylene Glycol Methyl Ether (DOWANOL DPM)	1.3
Water	82.0
	100.0

Use Dilution: 1 part/32parts water
Lit. Ref.: CS-408

SOURCE: Rohm and Haas Co.: Specialty Chemicals: Detergent
          Formulations for Industrial and Institutional Industry

METAL CLEANER, LOW-FOAM LIQUID-C

RAW MATERIALS   % Active

TRITON DF-20 Surfactant   2.0
Tetrapotassium Pyrophosphate (TKPP)   6.0
Potassium Hydroxide   12.0
Sodium Silicate   12.0
Water   68.0
                                                          100.0

METAL CLEANER, LOW-FOAM LIQUID-A

RAW MATERIALS   % Active

TRITON DF-20 Surfactant   2.0
Tetrapotassium Pyrophosphate (TKPP)   17.3
Potassium Hydroxide   14.1
Sodium Silicate   15.2
Water   51.4
                                                          100.0

METAL CLEANER, LOW-FOAM LIQUID-B

RAW MATERIALS   % Active

TRITON DF-20 Surfactant   2.0
Tetrapotassium Pyrophosphate (TKPP)   22.8
Potassium Hydroxide   9.0
Sodium Silicate   16.0
Water   50.2
                                                          100.0

Mixing Instructions:
    Add TRITON DF-20 Surfactant to water, then potassium hydroxide, and builders last.

Use Dilution: 1 to 2 oz./gal. water.

Approximate weight ratio SiO_2: Na_2O
    Formulation A--1.80
              B--2.50
              C--1.00

SOURCE: Rohm and Haas Co.: Specialty Chemicals: Detergent
        Formulations for Industrial and Institutional Industry:
        Lit. Ref.: CS-406

METAL CLEANER, MEDIUM DUTY

RAW MATERIALS	% By Weight
Sodium Metasilicate (Anhydrous)	34.0
Trisodium Phosphate (TSP)	34.0
TRITON X-100 Surfactant	5.0
TRITON QS-15 Surfactant	3.0
Soda Ash	24.0
	100.0

Use Dilution: 1 part/20 parts water.

Lit. Ref.: CS-427, CS-417

METAL CLEANER, SOAK TANK-A

RAW MATERIALS	% By Weight
TRITON QS-44 Surfactant (80%)	1.25
TRITON X-100 Surfactant	1.00
Sodium Hydroxide	40.00
Sodium Metasilicate (Anhydrous)	31.50
Soda Ash	26.25
	100.00

Use Dilution: 1 part/20 parts water

Lit. Ref: CS-410, CS-427, CS-439

METAL CLEANER, SOAK TANK-B

RAW MATERIALS	% By Weight
TRITON QS-30 Surfactant (90%)	2.20
Sodium Hydroxide	40.00
Sodium Metasilicate (Anhydrous)	31.50
Soda Ash	26.30
	100.00

Use Dilution: 1 part/20 parts water.

Lit. Ref.: CS-410, CS-427, CS-439

SOURCE: Rohm and Haas Co.: Specialty Chemicals: Detergent
Formulations for Industrial and Institutional Industry

METAL CLEANER--SPRAY ON TYPE

RAW MATERIALS   % By Weight

Butylcellosolve   20.0
IGEPAL CO-630   20.0
Kerosene   60.0
                                                            100.0

Manufacturing Procedure:
   1. Mix butylcellosolve and kerosene together.
   2. Add IGEPAL CO-630. Mix thoroughly.

Physical Properties:
   pH (as is)   7.5
   pH (1%)   6.4
   Viscosity   10 cps
   Specific Gravity   .96

SOURCE: GAF Corp.: Formulary: Prototype Formulation GAF 5655

METAL CLEANER FOR COPPER AND BRASS (METAL POLISH)

RAW MATERIALS   % By Weight

PHASE I:
Water   43.0
Cobratec 99   0.2
Vangel B   1.5
Super floss   15.0
Ammonium hydroxide   1.0
PHASE II:
Mineral spirits   30.0
Oleic acid   8.0
NINOL 11-CM   1.5

Mixing Procedure:
    Dissolve Cobratec in water. Slowly add Vangel B while agit-
ating at maximum available shear. Mix until smooth. Add Super
floss and ammonium hydroxide while under agitation. Combine
Phase II ingredients and mix until clear. Add Phase II to
Phase I while mixing and continue until uniform.

Appearance: creamy liquid

Use Instructions: Pour small amount on damp cloth. Clean article
    with moderate rubbing. Rinse with water. Dry and polish with
    a clean cloth.

SOURCE: Stepan Co.: Formulation No. 30

METAL CLEANER, SOLVENT-EMULSIFIER

RAW MATERIALS	% By Weight
TRITON X-45 Surfactant	12.0
Cresylic Acid	5.0
Kerosene	83.0
	100.0

Mixing Instructions:
    Add TRITON X-45 Surfactant to the kerosene then add cresylic acid slowly.

Directions for Use:
    Submerge metal parts in the solution. Agitate or scrub parts. After removing them, rinse with water and dry.

Lit. Ref.: CS-403

SOURCE: Rohm and Haas Co.: Specialty Chemicals: Detergent
        Formulations for Industrial and Institutional Use:
        Lit. Ref: CS-403

ACID METAL CLEANER

RAW MATERIALS	% By Weight
Water	86
Phosphoric acid (85%)	7
Ethylene glycol butyl ether	4
PLURAFAC D-25 surfactant	3

Use as is

SOURCE: BASF Corp.: Cleaning Formulary: Formulation #3301

ACID METAL CLEANER

RAW MATERIALS	% By Weight
Water	74
Citric acid	10
Ethylene glycol methyl ether	6
ICONOL TDA-10 surfactant	10

Use as is.

SOURCE: BASF Corp.: Cleaning Formulary: Formulation #3300

ACID METAL CLEANER

RAW MATERIALS   % By Weight

PHASE I:
Water   72.1
Kelzan   0.4
Veegum   0.8
PHASE II:
Hampene 100   0.9
MAKON 12   0.8
Phosphoric acid   15.0
Super floss   10.0

Mixing Procedure:
    Blend Veegum and Kelzan. Slowly add to the water while agit-
ating at maximum available shear. Continue mixing until smooth.
Add Phase II ingredients in order, mixing well after each addition until smooth.

Appearance: White pourable liquid

Use Instructions:
    Pour small amount on a damp cloth or sponge and clean article
with moderate rubbing. Rinse with water and dry.

SOURCE: Stepan Co.: Formulation No. 28

AIRCRAFT CLEANER

RAW MATERIALS   % By Weight

Sodium tripolyphosphate   30
Sodium carbonate   21
Linear alkyl aryl sulfonate (60%)   3
KLEARFAC AA-270 surfactant   1
Sodium metasilicate pentahydrate   45

Suggested use concentration: 1/4 to 3 oz. gallon of water

SOURCE: BASF Corp.: Formulation #3675

AIRCRAFT CLEANER

RAW MATERIALS   % By Weight

NEODOL 25-3S (60%)   33.4
Sodium xylene sulfonate (40%)   5.0
CYCLO SOL 63   45.9
SHELL SOL 340   9.7
Butyl OXITOL   5.0
Sodium nitrite   1.0

Properties:
    Viscosity, 73F, cps   14
    pH   9

SOURCE: Shell Chemical Co.: NEODOL Formulary: Formulations

ALKALINE METAL DEGREASING BATH

RAW MATERIALS	% By Weight
VARION EP AMVSF	5.0
Tetrapotassium Pyrophosphate (TKPP)	5.0
KOH	5.0
REWOPOL HV10 (Nonoynol-10)	5.0
Isopropanol (IPA)	3.0
1,2 Propylene Glycol	5.0
EDTA	10.0
Water	qs 100

Mixing Procedure:
    Dissolve TKPP, EDTA, and AMVSF in water. Add HV10 and IPA and Glycol and then the KOH.

SOURCE: Sherex: Industrial Formulation 1:05.5.2

ALKALI METAL DEGREASING BATH

RAW MATERIALS	% By Weight
VARION AMV	5.0
Potassium Pyrophosphate	5.0
Potassium Hydroxide	5.0
Isopropanol (IPA)	3.0
1,2 Propylene Glycol	5.0
Trilon B	2.0
Water	qs100

Mixing Procedure:
    Dissolve the Trilon B, Propylene Glycol, and phosphate into the water. Add the IPA and potassium hydroxide followed by the AMV.

SOURCE: Sherex: Industrial Formulation 1:05.5.2

ALKALINE SOAK TANK CLEANER FOR ALUMINUM

RAW MATERIALS	% By Weight
Sodium Metasilicate, Anhydrous	45
Sodium Tripolyphosphate	30
Sodium Bicarbonate	21
Sodium Alkyl Aryl Sulfonate	3
SURFONIC N-200	1

    Concentration: 4 oz./gal.
    Temperature: 160F

SOURCE: Texaco Chemical Co.: Suggested Formulation

ALKALINE SOAK TANK CLEANER FOR BRASS

RAW MATERIALS   % By Weight

Trisodium Phosphate   50
Sodium Metasilicate, Pentahydrate   30
Sodium Carbonate (soda ash)   13
SURFONIC N-95   7

    Concentration: 4-12 oz./gal.
    Temperature: 175-200F

SOURCE: Texaco Chemical Co.: Suggested Formulation

ALUMINUM BRIGHTENER/CLEANER

RAW MATERIALS   % By Weight

Water   45
Phosphoric acid (85%)   30
Ethylene glycol butyl ether   17
PLURAFAC D-25 surfactant   8

Use as is

SOURCE: BASF Corp.: Formulation #3375

ALUMINUM CLEANER AND BRIGHTENER(NON-HYDROFLUORIC TYPE)

RAW MATERIALS   % By Weight

Phosphoric acid, 75%   10.0
Hydrochloric acid   5.0
Ethylene glycol n-butyl ether   10.0
TRYCOL 5940 POE (6) Tridecyl Alcohol   4.0
TRYCOL 6964 POE (9) Nonylphenol   2.0
Water   69.0
                                                      100.0

Blending Procedure:
    Add the water to the blending tank. While mixing, add the
ingredients to the blending tank in the order listed. Mix
until uniform.
    Due to the high acid content follow proper precautions when
making and using this formula.

SOURCE: Emery Chemicals: Specialty Chemicals Formulary:
       Formulation 2886-083

ALKALINE METAL CLEANER-HIGH QUALITY

RAW MATERIALS   % By Weight

NEODOL 91-6   10.0
Sodium metasilicate, pentahydrate   7.0
Trisodium phosphate, anhydrous basis   2.0
Sodium hydroxide (50%)   3.0
EDTA   6.0
TRITON H-66   5.0
Water   to 100%

Properties:
    Viscosity, 73F, cps   28
    Phase coalescence temp., F   99
    pH   12.7

ALKALINE METAL CLEANER--GOOD QUALITY

RAW MATERIALS   % By Weight

NEODOL 91-6   7.5
Sodium metasilicate, pentahydrate   15.6
EDTA   1.0
TRITON H-66   8.0
Water   to 100%

Properties:
    Viscosity, 73F, cps   12
    Phase coalescence temp., F   135
    pH   13.2

Blending Procedure for Alkaline Metal Cleaners:
    Add builders last with vigorous mixing until homogeneous.

Recommended Dilution for Alkaline Metal Cleaners: 1-2 oz/gal.

SOURCE: Shell Chemical Co.: The NEODOL Formulary: Suggested
        Formulations

ALUMINUM CLEANER--ACID ALUMINUM BRIGHTENER

RAW MATERIALS   % By Weight

NEODOL 91-8   3.0
Phosphoric acid (85%)   3.0
Citric acid   4.0
Butyl OXITOL   4.0
Water   to 100%

Properties:
   Viscosity, 73F, cps   5
   Phase coalescence temp., F   >176
   pH   1.1

Blending Procedure for Aluminum Brightener:
   Dissolve acids in water. Add Butyl OXITOL and NEODOL. Stir
until clear solution is obtained.

ALUMINUM CLEANER--ALKALINE ALUMINUM CLEANER WITH PHOSPHATE*

RAW MATERIALS   % By Weight

NEODOL 91-6   3.0
C12 LAS (60%)   3.3
Sodium hydroxide (50%)   10.0
Sodium metasilicate, pentahydrate   3.0
Tetrapotassium pyrophosphate   2.0
TRITON H-66   7.0
Water   to 100%

Properties:
   Viscosity, 73F, cps   8
   Phase coalescence temp., F   >176
   pH   13.3

* Not intended for soak type operation

Blending Procedure for Aluminum Cleaner:
   Add builders last with vigorous mixing until homogeneous.

SOURCE: Shell Chemical Co.: The NEODOL Formulary: Suggested
       Formulation

ALUMINUM BRIGHTENER

RAW MATERIALS   % By Weight

Water	12
phosphoric acid (85%)	49
Igepal CO-630 surfactant	10
DOWANOL DPM glycol ether	25
o-dichlorobenzene	4

This product is effective in removing varnish and oxides from aluminum and metal surfaces.

DOWANOL DPM used as a coupling solvent for o-dichlorobenzene and phosphoric acid, also as a penetrant.

ALUMINUM CLEANER

RAW MATERIALS   % By Weight

A.	
stearic acid	4.0
Dow Corning Fluid 200, 350 ctk.	2.0
DOWANOL PM glycol ether	7.0
B.	
triethanolamine	0.7
water	24.0
C.	
Super-Floss powdered silica	14.0
METHOCEL 65 HG 4000 cellulose ether	0.1
D.	
water	48.2

1. Heat A to 60C.
2. Heat B to same temperature as A, but do not sustain this temperature for prolonged period.
3. When B approaches temperature of A, add B quickly to A with stirring until homogeneous dispersion results.
4. Heat D water to 90C and add C with stirring.
5. Add that mixture to A-B maintaining slow stirring until the entire mixture has cooled to room temperature.

SOURCE: Dow Chemical Co.: The Glycol Ethers Handbook: Suggested Formulation

ALUMINUM CLEANER (ACIDIC)

RAW MATERIALS	% By Weight
MIRANOL C2M-SF CONC.	5.0
Dowanol EB	6.0
Phosphoric Acid, 85%	38.0
Hydrofluoric Acid, 70%	8.0
Ethylenediamine Tetra Acetic Acid	1.0
Water	42.0

Although good soil removal qualities have been claimed for many acid cleaners, their principal functions are the removal of oxides and the brightening of aluminum. A two part process is almost unavoidable.

Thus, alkaline cleaners which are efficient soil removers, but are incapable of removing oxides are also used.

Bright aluminum parts may be cleaned by immersion in the following industrial cleaner.

ALUMINUM CLEANER I

RAW MATERIAL	% By Weight
MIRANOL C2M-SF CONC.	5.0-10.0
Sodium Metasilicate Pentahydrate	5.0- 5.0
Water	90.0-85.0

ALUMINUM CLEANER II

RAW MATERIALS	% By Weight
MIRANOL C2M-SF CONC.	3.5
Tetrapotassium Pyrophosphate	5.0
Sodium Metasilicate Pentahydrate	1.0
Triton X-100	1.5
Carbitol	3.0
Water	86.0

Note:
    In many cases, when aluminum parts dry between cleaning and rinsing a white film may appear, especially if silicates are used. This problem will not arise using this formulation.

SOURCE: Miranol Inc.: MIRANOL Products for Household/Industrial
        Applications: Suggested Formulations

CAUSTIC-GLUCONATE SOAK FORMULATION

RAW MATERIALS	% By Weight
Sodium Hydroxide	55-90
Sodium Gluconate	10-30
SURFONIC N-95	1-3
Sodium Alkyl Aryl Sulfonate	2-8

    Concentration: 8-48 oz./gal.
    Temperature: 140F to boiling

For electrolytic derusting, omit the wetting agent.

SOURCE: Texaco Chemical Co.: Suggested Formulation

COLD METAL CLEANER

RAW MATERIALS	% By Weight
VAROX 365	5.0
VARINE O	1.5
Butylene Glycol	5.0
REWOYL NXS 40	10.0
Isopropanol	2.5
VARAMIDE A2	2.5
Versene 100 (EDTA)	0.5
Water	73.0

Mixing Procedure:
    Stir into water, VAROX 365, NXS 40, Butylene Glycol, EDTA, IPA, VARINE O, and VARAMIDE A2

SOURCE: Sherex: Industrial Formulation 31:10.5

COLD METAL CLEANER

RAW MATERIALS	% By Weight
VAROX 365	5.0
VARINE O	1.5
Butylene Glycol	5.0
Reworyl NXS 40	10.0
Isopropanol	2.5
Versene 100	0.5
Water	73.0
VARAMIDE A2	2.5

Mixing Procedure:
    Add to water VAROX 365, NXS 40, Butylene Glycol, Versene 100, IPA, VARINE O and VARAMIDE A2. Stir all ingredients together.

SOURCE: Sherex: Industrial Formulation 33:10.5

COPPER CLEANER (PASTE)

RAW MATERIALS	% By Weight
Igepal CO-530 surfactant	9.5
DOWANOL TPM glycol ether	23.0
phosphoric acid (85%)	43.0
Celite diatomaceous earth	24.5

Blend Igepal with DOWANOL TPM. Add phosphoric acid and mix (exothermic). Add diatomaceous earth and mix until uniform.

DOWANOL TPM maintains the consistency of the blend, prevents drying out and aids stain removal.

SOURCE: Dow Chemical U.S.A.: The Glycol Ethers Handbook: Suggested Formulation

METAL CLEANER (SOAK CLEANER)

1. (Ferrous--Alkali non etching cleaner) (Not rust removers)

RAW MATERIALS	% By Weight
DOWANOL DPM glycol ether	5.00
trisodium phosphate	5.00
sodium orthosilicate	4.00
VERSENE 100 chelating agent	0.25
water	85.75

2. (Non-Ferrous)

	% By Weight
DOWANOL P-mix glycol ether	7.0
trisodium phosphate	5.0
sodium metasilicate	4.0
DOWFAX 2A-1 surfactant	0.2
water	83.8

For asphaltic type soils.

HEAVY-DUTY METAL CLEANER

RAW MATERIALS	% By Weight
water	80.1
sodium hydroxide	0.6
Triton X-102 surfactant	2.0
alkyl aryl sulfonic acid	2.9
tripotassium phosphate	2.2
sodium nitrite	0.2
Tamol SN dispersing agent	2.0
DOWANOL P-Mix glycol ether	10.0

SOURCE: Dow Chemical U.S.A.: The Glycol Ethers Handbook: Suggested Formulations

METAL CLEANER-ALKALINE WASH POWDER-WITH PHOSPHATE

RAW MATERIALS	% By Weight
NEODOL 91-6	5.6
NEODOL 91-2.5	2.4
Sodium metasilicate, anhydrous	35.0
Sodium tripolyphosphate	20.0
Sodium carbonate	22.0
Sodium hydroxide, flakes	15.0

METAL CLEANER--ALKALINE WASH POWDER--NON-PHOSPHATE POWDER

RAW MATERIALS	% By Weight
NEODOL 91-6	5.6
NEODOL 91-2.5	2.4
Sodium metasilicate, anhydrous basis	35.0
Sodium carbonate	37.0
Sodium hydroxide, flakes	20.0

METAL CLEANER--PAINT STRIPPER

RAW MATERIALS	% By Weight
NEODOL 25-3S (60%)	2.0
Sodium hydroxide (50%)	30.0
Sodium xylene sulfonate (40%)	3.0
Water	to 100%

Properties:

Viscosity, 73F, cps	8
Phase coalescence temp., F	>185
pH	13.3

SOURCE: Shell Chemical Co.: The NEODOL Formulary: Suggested
Formulations

METAL CLEANING-PAINT REMOVAL

RAW MATERIALS	% By Weight
AEROSOL C-61	20
Sodium Carbonate	40
Sodium Metasilicate	30
Sodium Hydroxide	10

Grind the powders together and add AEROSOL C-61 slowly while mixing.

METAL CLEANING--RUST REMOVAL

RAW MATERIALS	% By Weight
AEROSOL OS	10
Sodium Bifluoride	75
Sodium Tetraphosphate	10
Sodium Bisulfite	15

METAL CLEANING--ALKALINE

RAW MATERIALS	% By Weight
AEROSOL 22	10
Sodium Carbonate	40
Sodium Metasilicate	20
Sodium Hydroxide	10

Grind the powders together and add AEROSOL 22 slowly while mixing.

METAL CLEANING--ACID

RAW MATERIALS	% By Weight
Dowanol EB	6.0
Phosphoric Acid (85%)	7.0
AEROSOL 22	86.7
Water	86.7

SOURCE: Angus Chemical Corp.: Suggested Formulations

METAL CLEANER
Powder, for Spray Type

RAW MATERIALS	% By Weight
ANTAROX BL-240	5.0
Sodium hydroxide, flaked	40.0
Sodium metasilicate, anhydrous	40.0
Sodium carbonate (lt. density)	15.0
	100.0

Manufacturing Procedure:
1. Mix ANTAROX BL-240 and sodium carbonate together until uniform.
2. Add sodium hydroxide and sodium metasilicate to main batch. Mix thoroughly.

Physical Properties:
pH (1%)	12.7
Specific Gravity	1.01

SOURCE: GAF Corp.: Formulary: Prototype Formulation GAF 5607

SOAK TANK METAL CLEANER
Powder, for Aluminum

RAW MATERIALS	% By Weight
GAFAC RA-600	3.0
IGEPAL CO-710	2.0
Sodium bicarbonate	20.0
Sodium tripolyphosphate	25.0
Sodium metasilicate, anhydrous	50.0
	100.0

Manufacturing Procedure:
1. Mix surfactants together and add to sodium tripolyphosphate. Mix thoroughly to obtain a uniform powder.
2. Add sodium bicarbonate and sodium metasilicate. Mix well in a twin shell blender.

Physical Properties:
pH (1%)	12.1
Specific Gravity	1.00

SOURCE: GAF Corp.: Formulary: Prototype Formulation GAF 5623

RUST REMOVER(DIP PART CLEANERS)

RAW MATERIALS	% By Weight
I.	
phosphoric acid (85%)	30.0
DOWANOL DPM glycol ether	12.0
water	57.8
Triton X-100 surfactant	0.2
II.	
phosphoric acid (85%)	40.0
DOWANOL DPM glycol ether	10.0
water	49.8
Triton X-100 surfactant	0.2

Place metal part in the bath at 150-200F.
DOWANOL DPM provides penetration and controls rate of rust removal.

Miranol JEM conc. surfactant	2
DOWANOL DE glycol ether	5
potassium hydroxide (45%)	75
triethanolamine	12
water	6

Brush on cleaner. Soak for 10 minutes.

SOURCE: Dow Chemical U.S.A.: The Glycol Ethers Handbook: Suggested Formulations

SILVER CLEANING BATH

RAW MATERIALS	% By Weight
Water	80.0
Sulfuric Acid	10.0
VARION AMKSF	4.0
Igepal CO-620	5.0
4-Mercaptopropionic Acid	0.5
Perfume	0.5

SOURCE: Sherex: Industrial Formulation 53.05.5

SOAK TANK METAL CLEANER

RAW MATERIALS	% By Weight
Sodium tripolyphosphate	53
Sodium carbonate	10
PLURAFAC B-26 surfactant	2
Sodium metasilicate pentahydrate	35

Suggested use concentration: 4-12 oz. per gallon of water.

SOURCE: BASF Corp.: Formulation #3325

SOAK TANK METAL CLEANER
Powder, for Brass

RAW MATERIALS   % By Weight

GAFAC RA-600   5.0
IGEPAL CO-710   2.0
Sodium carbonate (lt. density)   13.0
Trisodium phosphate   50.0
Sodium metasilicate 5H2O   30.0
                                                           100.0

Manufacturing Procedure:
   Mix surfactants together and add to sodium carbonate. Mix
   thoroughly; then add other components.

Physical Properties:
   pH (1%)   12.2
   Specific Gravity   .90

SOURCE: GAF Corp.: Formulary: Prototype Formulation GAF 5602

SOAK TANK METAL CLEANER
Powder, for Zinc

RAW MATERIALS   % By Weight

GAFAC RA-600   5.0
Tall oil fatty acid soap   5.0
Sodium tripolyphosphate   90.0
                                                           100.0

Manufacturing Procedure:
   1. Mix GAFAC RA-600 and sodium tripolyphosphate thoroughly
     to obtain a uniform mixture.
   2. Add tall oil fatty acid soap to above mixture.

Physical Properties:
   pH (1%)   8.8
   Specific Gravity   .92

SOURCE: GAF Corp.: Formulary: Prototype Formulation GAF 5604

SOAK TANK METAL CLEANER
Powder, for Magnesium

RAW MATERIALS   % By Weight

GAFAC RA-600   5.0
IGEPAL CO-710   2.0
Tall oil fatty acid soap   5.0
Sodium tripolyphosphate   20.0
Sodium carbonate (lt. density)   18.0
Sodium hydroxide, flaked   20.0
Sodium metasilicate, anhydrous   30.0
                                                       100.0

Manufacturing Procedure:
   1. Mix sodium carbonate, GAFAC RA-600, IGEPAL CO-710, and
      tall oil fatty acid soap together until a uniform
      powdered mixture is obtained.
   2. Add rest of ingredients mix thoroughly.

Physical Properties:
   pH (1%)   12.8
   Specific Gravity   .98

SOURCE: GAF Corp.: Formulary: Prototype Formulation GAF 5605

SOAK TANK METAL CLEANER
Powder, for Copper

RAW MATERIALS   % By Weight

GAFAC RA-600   3.0
IGEPAL CO-710   2.0
Sodium metasilicate, anhydrous   50.0
Sodium carbonate (dense)   20.0
Sodium hydroxide, flaked   15.0
Sodium pyrophosphate   10.0
                                                       100.0

Manufacturing Procedure:
   1. Mix GAFAC RA-600, IGEPAL CO-710, and sodium carbonate
      together until a uniform powdered mixture is obtained.
   2. Add rest of dry ingredients to main batch. Mix well.

Physical Properties:
   pH (1%)   12.3
   Specific Gravity   .99

SOURCE: GAF Corp.: Formulary: Prototype Formulation GAF 5606

SPRAY METAL CLEANER

RAW MATERIALS   % By Weight

Sodium carbonate   45
ICONOL TDA-6 surfactant   10
Sodium metasilicate pentahydrate   30
Sodium hydroxide   15

Suggested use concentration: 1/4 to 3 oz. per gallon of water
Formulation #3351

SPRAY METAL CLEANER

RAW MATERIALS   % By Weight

Sodium tripolyphosphate   20
Sodium carbonate.   40
PLURAFAC D-25   5
Sodium metasilicate pentahydrate   35

Suggested use concentration: 1/4 to 3 oz. per gallon of water
Formulation #3352

STEAM METAL CLEANER

RAW MATERIALS   % By Weight

Water   79.5
KLEARFAC AA-270 surfactant   6
Sodium metasilicate pentahydrate   6
Tetrapotassium pyrophosphate   7
Triethanolamine   1.5

Suggested use concentration: 1/4 to 3 oz. per gallon of water.
Formulation #3353

STEAM METAL CLEANER

RAW MATERIALS   % By Weight

Sodium tripolyphosphate   52
Sodium carbonate   11
PLURAFAC B-25-5   2
Sodium metasilicate pentahydrate   35

Suggested use concentration: 1/4 to 3 oz per gallon of water.
Formulation #3354

SOURCE: BASF Corp.: Suggested Formulations

HIGH QUALITY METAL DE-OILING LIQUID CONCENTRATES (OIL SPILL/ RIG CLEANERS)

FOR HEAVY OIL

RAW MATERIALS	% By Weight
NEODOL 91-6	3.0
NEODOL 91-2.5	7.0
Sodium metasilicate, pentahydrate	7.0
EDTA	6.0
Sodium xylene sulfonate (40%)	22.5
Water, dye	to 100%

Properties:

Viscosity, 73F, cps	77
Phase coalescence temp., F	120
pH	13.5

GENERAL PURPOSE

RAW MATERIALS	% By Weight
NEODOL 91-6	5.0
NEODOL 91-2.5	5.0
Sodium metasilicate, pentahydrate	7.0
EDTA	6.0
CYCLO SOL 53	2.0
Sodium xylene sulfonate (40%)	17.5
Water, dye	to 100%

Properties:

Viscosity, 73F, cps	15
Phase coalescence temp., F	128
pH	12.7

FOR LIGHT OIL

RAW MATERIALS	% By Weight
NEODOL 91-6	7.0
NEODOL 91-2.5	3.0
Sodium metasilicate, pentahydrate	7.0
EDTA	6.0
Sodium xylene sulfonate (40%)	17.5
Water, dye	to 100%

Properties:

Viscosity, 73F, cps	9.0
Phase coalescence temp., F	>176
pH	12.7

SOURCE: Shell Chemical Co.: The NEODOL Formulary: Formulations

8. Oven Cleaners

AEROSOL OVEN CLEANER

RAW MATERIALS	% By Weight
water	46.0
VEEGUM T colloidal silicates	1.5
DOWANOL TPM glycol ether	20.0
IGEPAL CO-630 surfactant	1.0
sodium hydroxide 30%	12.0
DOWFAX 2A1 surfactant (45% sol'n)	15.0
isobutane	4.5

1. Add the chemicals in the order listed.
2. Add the chemicals slowly and with mixing.
3. Package in aerosol dispensers.
4. Shake well before using.

Can--Standard tin lined, low tin solder
Valve--Neoprene gasket, 70 durometer

DOWANOL TPM is used for the high temperature properties and coupling ability.

SOURCE: Dow Chemical USA: The Glycol Ethers Handbook: Suggested
        Formulation

OVEN CLEANER

RAW MATERIALS	% By Weight	CAS Registry Number
Potassium Hydroxide (45%)	85.0	1310-58-3
ESI-TERGE 330	15.0	Not Established
	100.0	

Procedure:
   Add as listed. Do not let temperature rise above 60C while mixing.

Specifications:
   % Active: 53.0%
   Viscosity: 3400 cps--LV #1 spindle @ 60 RPM to increase viscosity to about 4500 cps increase ESI-TERGE to 20% and decrease Potassium Hydroxide to 80%

Illustrative Formula. ESI suggests cutting solids in half.

SOURCE: Emulsion Systems, Inc.: Technical Service Bulletin
        Code 330-2

OVEN CLEANER

RAW MATERIALS	% By Weight
MIRANOL C2M-SF Conc.	3.0
ACRYSOL ICS-1	6.0
Potassium Hydroxide, 45%	22.0
Water	69.0

Procedure:
   Mix the water and ACRYSOL ICS-1 together. Slowly add the MIRANOL C2M-SF CONC. Under high agitation add the Potassium Hydroxide very slowly to insure uniformity.

SOURCE: Miranol Inc.: MIRANOL Products for Household/Industrial Applications: Suggested Formulation

OVEN CLEANER
Liquid Type

RAW MATERIALS	% By Weight
KELZAN	1.2
VEEGUM	.4
Water	80.4
Sodium hydroxide (50% active)	8.0
GAFAC RA-600	10.0

Manufacturing Procedure:
1. Heat water to 85-90C. Add VEEGUM and KELZAN. Stir for one hour.
2. Bring temperature to 55C. Add sodium hydroxide and GAFAC RA-600. Mix thoroughly.

Physical Properties:
   pH (as is): 12.8
   pH (1%): 10.9
   Viscosity: 830 cps
   Specific Gravity: 1.02

   SOURCE: GAF Corp.: Formulary: Prototype Formulation GAF 5751

OVEN CLEANER
(1) FOAM SPRAY TYPE

RAW MATERIALS	% By Weight	CAS Registry Number
Water	70.00	7732-18-5
ESI-TERGE 330	5.00	Not Established
ESI-TERGE 320	2.50	52276-83-2
ESI-TERGE S-10	2.50	61789-19-3
Potassium Hydroxide (KOH) 90%	20.00	1310-58-3

Procedure:
    Add water and ESI-TERGE's to a suitable tank. (A water jacketed kettle preferable). Add potassium hydroxide (KOH) very slowly so temperature does not exceed 50C.

Specifications:
    % Solids: 24-26
    % Active: 26-28
    pH: 14
    Viscosity: 10 cps max LV #2 spindle @ 60 RPM.
    Free KOH: 14-17%
(1) This cleaner forms a foam when used with proper foam trigger
    sprayer.

SOURCE: Emulsion Systems Inc.: Technical Service Bulletin 330-3

OVEN CLEANER

RAW MATERIALS	% By Weight
Water	67.7
Sodium Citrate	2.1
Sodium Metasilicate	4.2
50% NaOH	5.2
Monoethanolamine	4.2
Dowanol PM	3.1
VARION HC	3.1
50% NaOH	5.2
Dodecylbenzenesulfonic Acid (DDBSA)	3.1
Nonyl Phenol 9 mole Ethoxylate (IGEPAL CC)	2.1

Mixing Procedure:
    Additions are in the order listed. Stir in all ingredients. The solution should be clear through the addition of VARION HC. After the addition of DDBSA, the solution should be stirred until homogeneous, and then the CO630 can be added into the solution.

SOURCE: Sherex Chemical Co., Inc.: Industrial Formulation
        18: 01.7

CAUSTIC CLEANER
(OVEN CLEANER OR PAINT STRIPPER)

RAW MATERIALS	% Active	% By Weight
Water	88.0	74.5
ACRYSOL ICS-1 Thickener (30%)	1.5	5.0
TRITON X-100 Surfactant	0.5	0.5
Sodium Hydroxide (50%)	10.0	20.0

Properties: Brookfield Viscosity, cps.
  @ 0.5 rpm: 342,000
  @ 12 rpm: 18,200
  Appearance: Slightly hazy
Mixing Procedure:
  Add ACRYSOL ICS-1 Thickener to the water, then the surfactant
with adequate agitation. Add caustic solution slowly with high-
shear mixing.
Note: A flocculant precipitate may form upon adding the caustic
      solution. It disappears after a few minutes of agitation.
Use Dilution: As prepared
Lit. Ref.: CS-427/CS-504

SOURCE: Rohm and Haas Co.: Specialty Chemicals: Detergent
        Formulations for Industrial and Institutional Industry

CAUSTIC CLEANER
(OVEN CLEANER OR PAINT STRIPPER)

RAW MATERIALS	% Active	% By Weight
Water	77.75	53.7
ACRYSOL ICS-1 Thickener (30%)	1.75	5.8
TRITON X-100 Surfactant	0.5	0.5
Sodium Hydroxide (50%)	20.0	40.0

Properties: Brookfield Viscosity, cps.
  @ 0.5 rpm: 68,500
  @ 12 rpm: 8,200
  Appearance: Opaque
Mixing Procedure:
  Add Acrysol ICS-1 Thickener to the water, then the surfactant
with adequate agitation. Add caustic solution slowly with high-
shear mixing.
  Note: A flocculant precipitate may form upon adding the
caustic solution. It disappears after a few minutes of mixing.
Use Dilution: As prepared
Lit. Ref.: CS-427/CS-504

SOURCE: Rohm and Haas Co.: Specialty Chemicals: Detergent
    Formulations for Industrial and Institutional Industry

CAUSTIC CLEANER
(OVEN CLEANER OR PAINT STRIPPER)

RAW MATERIALS	% Active	% By Weight
Water	88.25	75.3
ACRYSOL ICS-1 Thickener (30%)	1.25	4.2
TRITON X-100 Surfactant	0.5	0.5
Sodium Hydroxide (50%)	10.0	20.0

Properties: Brookfield Viscosity, cps.
    @ 0.5 rpm: 40,700
    @ 12 rpm: 7,400
    Appearance: Slightly hazy
Mixing Procedure: Add ACRYSOL ICS-1 Thickener to the water, then
    the surfactant with adequate agitation. Add caustic solution
    slowly with high-shear mixing.
    Note: A flocculant precipitate may form upon adding the
          caustic solution. It disappears after a few minutes
          of agitation.
Use Dilution: As prepared
Lit. Ref.: CS-427/CS-504

SOURCE: Rohm and Haas Co.: Specialty Chemicals: Detergent
        Formulations for Industrial and Institutional Industry

CAUSTIC CLEANER
(OVEN CLEANER OR PAINT STRIPPER)

RAW MATERIAL	% Active	% By Weight
Water	78.0	54.5
ACRYSOL ICS-1 Thickener (30%)	1.5	5.0
TRITON X-100 Surfactant	0.5	0.5
Sodium Hydroxide (50%)	20.0	40.0

Properties: Brookfield Viscosity, cps.
    @ 0.5 rpm: 24,000
    @ 12 rpm: 2,500
    Appearance: Opaque
Mixing Procedure: Add ACRYSOL ICS-1 Thickener to the water,
    then the surfactant with adequate agitation. Add caustic
    solution slowly with high-shear mixing.
    Note: A flocculant precipitate may form upon adding the
          caustic solution. It disappears after a few minutes
          of agitation.
Use Dilution: As prepared
Lit. Ref.: CS-427/CS-504

SOURCE: Rohm and Haas Co.: Specialty Chemicals: Detergent
        Formulations for Industrial and Institutional Industry

9. Rinse Additives and Aids

RINSE ADDITIVES BIODEGRADABLE, MACHINE DISHWASHING

RAW MATERIALS   % By Weight

A.
TRITON DF-16	30.0
Water	48.85
TRITON X-45 Surfactant	10.0
TRITON DF-12 Surfactant	10.0
Hyamine 3500	0.25
Sarkosyl NL-30	0.90

    Cloud Point C, Conc.: 64
    Specific Gravity at 25C: 1.02
    Viscosity cps, 25C: 70

B:
TRITON DF-16	50.0
Water	40.0
Isopropyl Alcohol	10.0

C.
TRITON DF-16	35.0
Water	35.0
TRITON X-45 Surfactant	15.0
Isopropyl Alcohol	10.0
Phosphoric Acid	5.0

    Cloud Point C, Conc.: 55
    Specific Gravity at 25C: 1.03
    Viscosity cps, 25C: 100

D:
TRITON DF-16	42.0
Water	58.0
D&C Red #33 Dye*	0.002

  * Formulation D conforms to Federal Spec P-R-1272C.

Use Dilution: 100 ppm in final rinse

SOURCE: Rohm and Haas Co.: Specialty Chemicals: Detergent
       Formulations for Industrial and Institutional Industry:
       Lit. Ref: CS-403/CS-415/CS-448

RINSE ADDITIVES, MACHINE DISHWASHING

RAW MATERIALS	% By Weight
TRITON CF-10 Surfactant	70-100
Water	30-0
Cloud F.: 140	
Point C.: 60	
TRITON CF-10 Surfactant	50
Water	30
Isopropyl Alcohol	20
Cloud F.: 140	
Point C.: 60	
TRITON CF-10 Surfactant	50
Water	40
Glycolic Acid	10
Cloud F.: 126	
Point C.: 52	
TRITON CF-10 Surfactant	30
Water	32
TRITON X-45 Surfactant	20
Isopropyl Alcohol	18
Cloud F.: 140	
Point C.: 60	
TRITON CF-10 Surfactant	48
Water	45
Sodium Lauroyl-sarcosinate (Sarkosyl NL-30)	7
Cloud F.: 140	
Point C.: 60	
TRITON CF-10 Surfactant	47.5
Water	42.5
TRITON H-66 Surfactant (50%)	10
Cloud F: 133	
Point C: 56	

Directions for Use: 100 ppm in final rinse

SOURCE: Rohm and Haas Co.: Specialty Chemicals: Detergent
       Formulations for Industrial and Institutional Industry:
       Lit. Ref: CS-403, CS-422, CS-433, CS-436

RINSE ADDITIVES, MACHINE DISHWASHING

RAW MATERIALS   % By Weight

Type A--Acidic for Hard Water:

TRITON CF-87 Surfactant (90%)	56.00
Glycolic Acid (70%)	9.50
Isopropyl Alcohol	10.00
Water	24.50
	100.00

Type B--Neutral for Soft Water:

TRITON CF-87 Surfactant (90%)	50.50
TRITON X-45 Surfactant	4.50
Sodium Hydroxide (5%)	2.25
Water	42.75
	100.00

Use Dilution: 100 ppm in final rinse

SOURCE: Rohm and Haas Co.: Specialty Chemicals: Detergent
        Formulations For Industrial and Institutional Industry:
        Lit. Ref: CS-403, CS-423

RINSE ADDITIVES, MACHINE DISHWASHING

RAW MATERIALS	% By Weight
TRITON CF-87 (90%) Surfactant	66.7
Water	33.15
Sodium Hydroxide	0.15
Cloud Point C: 72	
Viscosity cps: 235	
TRITON CF-87 (90%) Surfactant	66.7
Water	27.50
Phospnoric Acid (86%)	5.8
Cloud Point C: 65	
Viscosity cps: 290	
TRITON CF-87 (90%) Surfactant	55.6
Water	38.28
Sodium Hydroxide	0.12
Isopropyl Alcohol	6.0
Cloud Point C: 68	
Viscosity cps: 80	
TRITON CF-87 (90%) Surfactant	55.6
Water	32.60
Phosphoric Acid (86%)	5.8
Isopropyl Alcohol	6.0
Cloud Point C: 61	
Viscosity cps: 110	
TRITON CF-87 (90%) Surfactant	44.4
Water	49.50
Sodium Hydroxide	0.10
Isopropyl alconol	6.0
Cloud Point C: 62	
Viscosity cps: 52	
TRITON CF-87 (90%) Surfactant	31.1
Water	62.83
Sodium Hydroxide	0.07
Isopropyl Alcohol	6.0
Cloud Point C: 54	
Viscosity cps: 15	

Directions for use: 100 ppm in final rinse

SOURCE: Rohm and Haas Co.: Specialty Chemicals: Detergent
Formulations for Industrial and Institutional Industry:
(6781E/270Z)

RINSE AID CONCENTRATES

RAW MATERIALS	% By Weight
#1:	
MAZER MACOL 24 or 30	82
Isopropanol	6
Propylene Glycol	6
Water	6
#2:	
MAZER MACOL 24 or 30	75
Isopropanol	7
Propylene Glycol	6
Water	12

Both of the before mentioned formulas are for end use at 20-100 ppm in water temperatures between 105F-130F.

SOURCE: Mazer Chemicals, Inc.: Household/Industrial T-20B:
        Formulations 11-#1, 11-#2

DISHWASHER RINSE AID

RAW MATERIALS	% By Weight
A.	
SURFONIC LF-17	70-100
Water	0-30
B.	
SURFONIC LF-17	50
Water	30
Additive(s)	
Isopropyl alcohol	20
C.	
SURFONIC LF-17	50
Water	30
Additive(s)	
Isopropyl alcohol	17
Propylene glycol	3

SOURCE: Texaco Chemical Co.: Suggested Formulations

RINSE AID--ECONOMICAL TYPE

RAW MATERIALS   % By Weight

ANTAROX BL-240   26.0
IGEPAL CO-430   26.0
Isopropanol   20.0
Water   28.0
                                                             100.0

Perfume and colorants added, as desired, replacing water.

Manufacturing Procedure:
    1. Dissolve ANTAROX BL-240 in water.
    2. Add isopropanol.
    3. Add IGEPAL CO-430. Mix thoroughly.

Physical Properties:
    pH (as is)   8.2
    pH (1%)   8.4
    Viscosity   30 cps
    Specific Gravity   .99

SOURCE: GAF Corp.: Formulary: Prototype Formulation GAF 5240

RINSE AID--LOW DENSITY TYPE

RAW MATERIALS   % By Weight

ANTAROX BL-225   70.0
Isopropanol   20.0
Water   10.0
                                                             100.0

Perfume and colorants added, as desired, replacing water.

Manufacturing Procedure:
    1. Dissolve ANTAROX BL-225 in water. Mix thoroughly.
    2. Add isopropanol.

Physical Properties:
    pH (as is)   6.1
    pH (1%)   8.3
    Viscosity   30 cps
    Specific Gravity   .99

SOURCE: GAF Corp.: Formulary: Prototype Formulation GAF 5241

RINSE AID--HIGH DENSITY TYPE

RAW MATERIALS	% By Weight
ANTAROX BL-240	50.0
Urea	9.0
Water	41.0
	100.0

Perfumes and colorants added, as desired, replacing water.

Manufacturing Procedure:
1. Dissolve urea in water.
2. Add ANTAROX BL-240 to water/urea solution. Mix thoroughly.

Physical Properties:
pH (as is)	7.5
pH (1%)	8.1
Viscosity	50 cps
Specific Gravity	1.01

SOURCE: GAF Corp.: Formulary: Prototype Formulation GAF 5242

DISHWASHER RINSE AIDS

RAW MATERIALS	% By Weight
HAMPOSYL L-30	7
Triton CF-21	48
Water	45
HAMPOSYL L-30	7
Triton CF-10	23
Triton CF-21	23
Water	47
HAMPOSYL L-30	5
Triton CF-54	50
Water	45

Dishwasher Rinse Aids are formulated from the HAMPOSYL sar-cosinates and low-foam nonionics. The sarcosinate surfactant raises the cloud point and solubility of the nonionic and increases wetting action without contributing added foam.

SOURCE: W.R. Grace & Co.: HAMPOSYL Surfactants: Suggested Formulations

RINSE AID FOR DISHWASH

RAW MATERIALS	% By Weight
VARIDRI 40	4.0
AROSULF 42 PE10	6.0
Isopropanol (IPA)	12.0
Citric Acid	10.0
Water	68.0

Mixing Procedure:
    Mix in Citric Acid to water. Add to solution, IPA, VARIDRI
40, 42PE 10. Liquid should be clear.

SOURCE: Sherex: Formulation 29:01.2.3

RINSE AID WITH MAKON NF-5

RAW MATERIALS	% By Weight
A:	
MAKON NF-5	50
Isopropyl alcohol	6
Soft water	Q.S.
B:	
MAKON NF-5	50
STEPANATE X	15
Soft water	Q.S.

Mixing Procedure:
    Charge tank with soft water. Add IPA or STEPANATE X and mix.
Add MAKON NF-5 last while under agitation.

Properties:
    Appearance   clear liquid
    Actives, %   56
    Cloud point (A), F   48
    Cloud point (B), F   76

Use Instructions:
    Concentration to give 25-100 ppm MAKON NF-5

Performance:
    Low foaming, excellent wetting, rapid drainage with no break
in the water film, good dispersability at use temperatures.
Leaves surfaces sparkling and spot-free.

SOURCE: Stepan Co.: Formulation No. 7

RINSE AIDS
For High Temperature Machines

RAW MATERIALS	% By Weight
No.1:	
PLURONIC L10 Surfactant	20
Water	to 100
Compatibility Limit, F	123
No. 2:	
PLURONIC L10 Surfactant	20
PETRO ULF	3
Water	to 100
Compatibility Limit, F	150
No. 3:	
PLURONIC L10 Surfactant	40
Water	to 100
Compatibility Limit, F	132
No. 4:	
PLURONIC L10 Surfactant	40
MONAWET MM-80	3
Water	to 100
Compatibility Limit, F	167
No. 5:	
PLURONIC L62 Surfactant	20
PETRO 22	3
Water	to 100
Compatibility Limit, F	132
No. 6:	
PLURONIC L62 Surfactant	20
MONAWET MM-80	1.5
Water	to 100
Compatibility Limit, F	127
No. 7:	
PLURONIC L62 Surfactant	40
PETRO 22	5
Water	to 100
Compatibility Limit, F	122
No. 8:	
PLURONIC L62 Surfactant	40
MONAWET MM-80	1
Isopropyl alcohol	5
Water	to 100
Compatibility Limit, F	120

RINSE AIDS
For High Temperature Machines

RAW MATERIALS   % By Weight

No. 9:
PLURONIC L62D Surfactant   20
PETRO ULF   3
Water   to 100
  Compatibility Limit, F   141

No. 10:
PLURONIC L62D Surfactant   20
MONAWET MM-80   3
Water   to 100
  Compatibility Limit, F   168

No. 11:
PLURONIC L62D Surfactant   40
PETRO ULF   6
Water   to 100
  Compatibility Limit, F   142

No. 12:
PLURONIC L62D Surfactant   40
MONAWET MM-80   3
Water   to 100
  Compatibility Limit, F   130

No. 13:
PLURAFAC RA-20 Surfactant   20
PETRO ULF   3
Water   to 100
  Compatibility Limit, F   143

No. 14:
PLURAFAC RA-20 Surfactant   20
MONAWET MM-80   3
Water   to 100
  Compatibility Limit, F   154

No. 15:
PLURAFAC RA-20 Surfactant   40
PETRO ULF   3
Water   to 100
  Compatibility Limit, F   140

No. 16:
PLURAFAC RA-20 Surfactant   40
MONAWET MM-80   3
Water   to 100
  Compatibility Limit, F   152

<u>RINSE AIDS</u>
For High Temperature Machines

RAW MATERIALS   % By Weight

No. 17:
PLURAFAC RA-30 Surfactant   20
PETRO ULF   0.7
Water   to 100
   Compatibility Limit, F   128

No. 18:
PLURAFAC RA-30 Surfactant   20
MONAWET MM-80   0.5
Water   to 100
   Compatibility Limit, F   132

No. 19:
PLURAFAC RA-30 Surfactant   40
PETRO ULF   1
Water   to 100
   Compatibility Limit, F   122

No. 20:
PLURAFAC RA-30 Surfactant   40
MONAWET MM-80   1
Water   to 100
   Compatibility Limit, F   125

For Low Temperature Machines

No. 21:
PLURONIC L61 Surfactant   20
MONAWET MM-80   2
Water   to 100
   Compatibility Limit, F   144

No. 22:
PLURONIC L61 Surfactant   20
Petro 22 (50%)   4
Water   to 100
   Compatibility Limit, F   120

No. 23:
PLURONIC L61 Surfactant   40
Monawet MM-80   1.5
Isopropyl alcohol   5
Water   to 100
   Compatibility Limit, F   140

No. 24:
PLURONIC L61 Surfactant   40
PETRO ULF   4
Isopropyl alcohol   5
Water   to 100
   Compatibility Limit, F   120

RINSE AIDS
For Low Temperature Machines

RAW MATERIALS	% By Weight

No. 25:

PLURONIC 25R2 Surfactant	20
PETRO ULF	6
Water	to 100
Compatibility Limit, F	114

No. 26:

PLURONIC 25R2 Surfactant	20
MONAWET MM-80	3
Water	to 100
Compatibility Limit, F	135

No. 27:

PLURONIC 25R2 Surfactant	40
PETRO ULF	6
Water	to 100
Compatibility Limit, F	104

No. 28:

PLURONIC 25R2 Surfactant	40
MONAWET MM-80	6
Water	to 100
Compatibility Limit, F	118

No. 29:

PLURAFAC RA-40 Surfactant	20
MONAWET MM-80	6
Water	to 100
Compatibility Limit, F	124

No. 30:

PLURAFAC RA-40 Surfactant	20
WITCOLATE D-510	6
Water	to 100
Compatibility Limit, F	116

No. 31:

PLURAFAC RA-40 Surfactant	40
MONAWET MM-80	6
Water	to 100
Compatibility Limit, F	123

No. 32:

PLURAFAC RA-40 Surfactant	40
WITCOLATE D-510	6
Water	to 100
Compatibility Limit, F	109

RINSE AIDS
For Low Temperature Machines

RAW MATERIALS	% By Weight
No. 33:	
INDUSTROL N3 Surfactant	20
MONAWET MM-80	2
Water	78
Compatibility Limit, F	123
No. 34:	
INDUSTROL N3 Surfactant	20
PETRO ULF	6
Water	74
Compatibility Limit, F	124
No. 35:	
INDUSTROL N3 Surfactant	40
PETRO ULF	6
Water	54
Compatibility Limit, F	116
No. 36:	
INDUSTROL N3 Surfactant	40
MONAWET MM-80	6
Water	54
Compatibility Limit, F	121

For Low Actives Formulations (10 wt.% surfactants)

PLURONIC L10:	
No. 37:	
PLURONIC Surfactant	10
MONAWET MM-80	1
Water	to 100
Compatibility Limit, F	140
No. 38:	
PLURONIC Surfactant	10
CALSOFT T-60	1.3
Water	to 100
Compatibility Limit, F	140
PLURONIC L62D:	
No. 39:	
PLURONIC Surfactant	10
MONAWET MM-80	1
Water	to 100
Compatibility Limit, F	140
No. 40:	
PLURONIC Surfactant	10
CALSOFT T-60	1.4
Water	to 100
Compatibility Limit, F	140

RINSE AIDS
For Low Actives Formulations (10 wt.% Surfactants)

RAW MATERIALS	% By Weight
PLURONIC L61	
No. 41:	
PLURONIC Surfactant	10
MONAWET MM-80	1
Water	to 100
Compatibility Limit, F	120
No. 42:	
PLURONIC Surfactant	10
CALSOFT T-60	2
Water	to 100
Compatibility Limit, F	120
PLURONIC 25R2	
No. 43:	
PLURONIC Surfactant	10
CALSOFT T-60	2.6
Water	to 100
Compatibility Limit, F	120
No. 44:	
PLURONIC Surfactant	10
NINATE 411	3
Water	to 100
Compatibility Limit, F	141
PLURAFAC RA-20	
No. 45:	
PLURAFAC Surfactant	10
PETRO ULF	1
Water	to 100
Compatibility Limit, F	134
No. 46:	
PLURAFAC Surfactant	10
MONAWET MM-80	0.5
Water	to 100
Compatibility Limit, F	135
PLURAFAC RA-30	
No. 47:	
PLURAFAC Surfactant	10
PETRO ULF	1.5
Water	to 100
Compatibility Limit, F	138
No. 48:	
PLURAFAC Surfactant	10
Monawet MM-80	1
Water	to 100
Compatibility Limit, F	146

RINSE AIDS
For Low Actives Formulations (10 wt.% Surfactants)

RAW MATERIALS	% By Weight
PLURAFAC RA-40	
No. 49:	
Surfactant	10
CALSOFT T-60	3
Water	to 100
Compatibility Limit, F	120
No. 50:	
Surfactant	10
NINATE 411	3
Water	to 100
Compatibility Limit, F	136
INDUSTROL N3	
NO. 51:	
Surfactant	10
CALSOFT T-60	1.5
Water	to 100
Compatibility Limit, F	88.5
No. 52:	
Surfactant	10
PETRO ULF	4.5
Water	85.5
Compatibility Limit, F	to 120

For Hard Water Formulations

	% By Weight
PLURONIC L10	
No. 53:	
PLURONIC Surfactant	20
PETRO ULF	3
Hydroxyacetic Acid (70%)	9
Water	to 100
Compatibility Limit, F	150
No. 54:	
PLURONIC Surfactant	40
Hydroxyacetic acid (70%)	9
Water	to 100
Compatibility Limit, F	134
PLURONIC L62D	
No. 55:	
PLURONIC Surfactant	20
PETRO ULF	1
Hydroxyacetic acid (70%)	9
Water	to 100
Compatibility Limit, F	155

RINSE AIDS
For Hard Water Formulations

RAW MATERIALS	% By Weight

PLURONIC L62D

No. 56:

PLURONIC Surfactant	40
PETRO ULF	3
Hydroxyacetic acid (70%)	9
Water	to 100
Compatibility Limit, F	164

PLURONIC RA-20

No. 57:

Surfactant	20
PETRO ULF	3
Hydroxyacetic acid (70%)	9
Water	to 100
Compatibility Limit, F	140

No. 58:

Surfactant	40
MONAWET MM-80	3
Hydroxyacetic acid (70%)	9
Water	to 100
Compatibility Limit, F	134

INDUSTROL N3

No. 59:

Surfactant	20
PETRO ULF	9
Hydroxyacetic acid (70%)	9
Water	to 100
Compatibility Limit, F	122

No. 60:

Surfactant	40
PETRO ULF	11
Hydroxyacetic acid (70%)	9
Water	to 100
Compatibility Limit, F	120

RINSE AIDS
For Hard Water Formulations

RAW MATERIALS   % By Weight

PLURONIC L61

No. 61:
Surfactant   20
PETRO ULF   6
Hydroxyacetic acid (70%)   9
Water   to 100
   Compatibility Limit, F   120

No. 62:
Surfactant   40
PETRO ULF   6
Isopropyl Alcohol   5
Hydroxyacetic acid (70%)   9
Water   to 100
   Compatibility Limit, F   126

PLURAFAC RA-40

No. 63:
Surfactant   20
MONAWET MM-80   6
Hydroxyacetic acid (70%)   9
Water   to 100
   Compatibility Limit, F   136

No. 64:
Surfactant   40
MONAWET MM-80   6
Hydroxyacetic acid (70%)   9
Water   to 100
   Compatibility Limit, F   137

SOURCE: BASF Corp.: Rinse Aid Formulary: Formulation No. 1 to
        No. 64

10. Rug, Carpet and Upholstery Cleaners and Shampoos

ANTISTATIC CARPET SHAMPOO

RAW MATERIALS   % By Weight

A--16% Resin Solution:
Water   79.5
Ammonium Hydroxide, 26 Be   4.5
SMA 2625 resin   16.0
                                                            100.0

B--Finished Product:
Water   36.75
Sodium Tripolyphosphate (STPP)   3.00
Citric acid   0.25
Disodium laureth sulfosuccinate   15.00
EMERSAL 6400 Sodium Lauryl Sulfate   20.00
EMID 6515 Coconut Diethanolamide   4.00
TRYMEEN 6606 POE (15) Tallow Amine   1.00
Resin solution (Part A)   20.00
Fragrance, optical brightener   as desired
                                                            100.00

Blending Procedure:
A: Add the water to the blending tank. While mixing, add the
     ingredients in the order listed. Bring the mixture to 125-
     135F and mix until clear.
B: Add the water to the blending tank. While mixing, add the
     STPP and citric acid. Make sure these ingredients are com-
     pletely dissolved before proceeding. Add the remaining
     ingredients in the order listed.

Use Dilution:
     Dilute one part of the shampoo to 10-15 parts water.

SOURCE: Emery Chemicals: Specialty Chemicals Formulary:
            Formulation 2878-40

CARPET CLEANER

RAW MATERIALS	% By Weight
Water	56.0
Trisodium Phosphate	15.0
Tetrapotassium Pyrophosphate	10.0
MAZER MACOL 25	7.0
MAZER MACOL 41	10.0
MAZER MAZAWET 77	1.25
EDTA-Tetrasodium Salt	0.5
Tinopal 5 BM	0.25

SOURCE: Mazer Chemicals, Inc.: Household/Industrial T-20B: 24

CARPET CLEANER(FOAM TYPE)

RAW MATERIALS	% By Weight
Water	85
Isopropyl alcohol	8
PLURAFAC C-17 surfactant	5
Sodium tripolyphosphate	2

Suggested use concentration: 1 gallon in 4 gallons of water

SOURCE: BASF Corp.: Cleaning Formulary: Formulation #3751

CARPET CLEANER(SCRUB TYPE)

RAW MATERIALS	% By Weight
Water	84
Triethanolamine	6
PLURAFAC D-25 surfactant	3
Coconut diethanolamide	7

Suggested use concentration: 1 gallon in 4 gallons of water

SOURCE: BASF Corp.: Cleaning Formulary: Formulation #3750

CARPET EXTRACTION CLEANER

RAW MATERIALS	% By Weight
MIRANOL JEM CONC.	5.0
Dowanol EB	2.0
Tetrapotassium Pyrophosphate	4.0
Water	89.0

SOURCE: Miranol Inc.: MIRANOL Products for Household/Industrial
Applications: Suggested Formulation

CARPET EXTRACTOR CLEANER

RAW MATERIALS	% By Weight	CAS Registry Number
Water	87.9	7732-18-5
Tetra Potassium Pyrophosphate	4.4	7720-88-5
Sodium Carbonate	2.2	497-19-8
EDTA	2.2	Not Established
ESI-TERGE RT-61	3.3	Not Established
	100.0	

Specifications:
% Solids	12.1
% Active	12.1
pH	11.1
Viscosity	2.5 cps.

SOURCE: Emulsion Systems Inc.: Technical Service Bulletin
Code RT-61-2

CARPET STEAM CLEANER

RAW MATERIALS	% By Weight	CAS Registry Number
Water	78.5	7732-18-5
Sodium Tripolyphosphate	4.0	7758-29-4
KOH 45%	7.0	1310-58-3
ESI-TERGE 330	10.0	Not Established
ESI-TERGE T-60	0.5	27323-41-7
	100.0	

Procedure:
    Add salts to water and dissolve. Add other ingredients in
order mentioned.
Specifications:
% Solids	17.65
% Active	17.45
pH	8.0-8.5
Viscosity	Water Like

SOURCE: Emulsion Systems Inc.: Technical Service Bulletin
Code 330-5

CARPET SHAMPOO

RAW MATERIALS   % By Weight

A--16% Resin Solution
Water   79.5
Ammonium hydroxide, 26 Be   4.5
SMA 2625 resin   16.0
                                                                  100.0

B--Finished Product
Water   31.0
EMERSAL 6400 Sodium Lauryl Sulfate   40.0
EMID 6515 Coconut Diethanolamide   4.0
Resin solution (Part A)   25.0
Fragrance, optical brightener   as desired
                                                                  100.0

Blending Procedure:
A: Add the water to the blending tank. While mixing, add the
   ingredients in the order listed. Bring the mixture to 125-
   135F and mix until clear.
B: Add the water to the blending tank. While mixing, add the
   ingredients in the order listed.

Use Dilution:
   Dilute one part of the shampoo to 10-15 parts water.

SOURCE: Emery Chemicals: Specialty Chemicals Formulary:
        Formulation 2878-041

CARPET SHAMPOO, AEROSOL

RAW MATERIALS	% By Weight
PRIMAPEL C-93 Polymer (25%)	18.0
Water	57.4
Dipropylene Glycol Methyl Ether (Dowanol DPM)	1.6
Sodium Lauryl Sulfate (29%) (Sipon LSB)	15.5
Isobutane	7.5
Perfume	As Required
Use Level--As Prepared	100.0

Properties: pH   9.6
             Stability @ 50C   Stable
             Resistance to Freeze/Thaw Cycling   Acceptable

SOURCE: Rohm and Haas Co.: Specialty Chemicals: Detergent
        Formulations for Industrial and Institutional Industry:
        Lit. Ref.: CS-502

CARPET SHAMPOO, SCRUBBING MACHINE

RAW MATERIALS	% By Weight
Water	23.3
PRIMAPEL C-93 Polymer (25%)	40.0
Dipropylene Glycol Methyl Ether (Dowanol DPM)	1.0
Ammonium Lauryl Sulfate (Sipon L-22--28%)	35.7
	100.0

Use Level: 1 part/10-40 parts water

Performance (2% solids)
    % Cleaning   63
    % Retardancy
        Initial   17
        Resoil   36

SOURCE: Rohm and Haas Co.: Specialty Chemicals: Detergent
        Formulations for Industrial and Institutional Industry:
        Lit. Ref.: CS-502

CARPET SHAMPOO, SCRUBBING MACHINE

RAW MATERIALS   % By Weight

Water   25.7
PRIMAPEL C-93 Polymer (25%)   40.0
Dipropylene Glycol Methyl Ether (Dowanol DPM)   1.0
Ammonium Lauryl Sulfate (Conoco Sulfate A--30%)   33.3
                                                                    100.0

Use Level: 1 part/10-40 parts water

After Storage 24 hrs./55C.
   pH   9.9
   cps/23C.   250
   4 weeks/55C.
   cps/23C.   224
   3 Freeze-Thaw Cycles
   cps/23C.   236

Performance (2% Solids)
   % Cleaning   56
   % Retardancy
     Initial   19
     Resoil   36

SOURCE: Rohm and Haas Co.: Specialty Chemicals: Detergent
       Formulations for Industrial and Institutional Industry:
       Lit. Ref.: CS-502

CARPET SHAMPOO, STEAM CLEANER

RAW MATERIALS   % By Weight

TRITON X-114 Surfactant   4.0
Linear Alkyl naphthalenesulfonate (Petro 22)   3.5
Sodium Tripolyphosphate (STPP)   5.0
Trisodium Phosphate (TSP)   2.0
Sodium Silicate (2.4 SiO2/Na2O--38%)   24.0
Water   61.5
                                                                   100.0

Use Dilution: 1 oz. per gal. water

SOURCE: Rohm and Haas Co.: Specialty Chemicals: Detergent
       Formulations for Industrial and Institutional Industry:
       Lit. Ref: CS-409

CARPET SHAMPOO, STEAM CLEANING

RAW MATERIALS	% By Weight
PRIMAPEL C-93 Polymer (25%)	56.0
Water	38.0
Sodium Alkyl Naphthalenesulfonate (100%)	6.0
Perfume	As Required
	100.0

Properties:
    Viscosity, cps @ 22C (72F)   110
    pH   10.2
    Stability @ 50C   Stable at least 30 days
    Resistance to Freeze/Tnaw Cycling   Acceptable

Use Dilution: 1 part/10-40 parts water

SOURCE: Rohm and Haas Co.: Specialty Chemicals: Detergent
        Formulations for Industrial and Institutional Industry:
        Lit. Ref.: CS-502

CARPET SHAMPOO, STEAM EXTRACTOR

RAW MATERIALS	% By Weight
Water	50.0
PRIMAPEL C-93 Polymer (25%)	40.0
TRITON N-101 Surfactant	10.0
	100.0

Use Level: 1 part/40 parts water

	% Cleaning	Soil Retardancy Initial	Resoil
Hot Water	56	32	10
TRITON N-101	72	0	0
Formulation	78	24	20

Note: Carpet cleaning tests using the above formulation show
      that PRIMAPEL C-93 Polymer improves the detergency and
      soil retardancy of steam extractor
      formulations based on nonionic surfactants.

SOURCE: Rohm and Haas Co.: Specialty Chemicals: Detergent
        Formulations for Industrial and Institutional Industry:
        Lit. Ref.: CS-408/CS-502

CARPET SHAMPOO TREATMENT
SOIL RETARDANT AND ANTISTAT
CLEANING TECHNIQUE--PUMP SPRAY

RAW MATERIALS	% By Weight
PRIMAPEL C-93 Polymer (25%)	12.0
Dipropylene Glycol Methyl Ether (Dowanol DPM)	0.1
Fluorad FC-129 (1%)	1.0
Nopcostat HS	0.5
Water	86.4
Perfume	As Required
	100.0

Use Level: As Supplied

Properties:

Appearance	Clear Solution
Stability (30 days @ 50C.)	Stable
Freeze-Thaw Resistance	3 Cycles
Soil Retardancy Improvement	61%
Antistat Performance Improvement	68%

SOURCE: Rohm and Haas Co.: Specialty Chemicals: Detergent
        Formulations for Industrial and Institutional Industry:
        Lit. Ref.: CS-502

CARPET SHAMPOO TREATMENT
SOIL RETARDANT AND ANTISTAT
CLEANING TECHNIQUE--AEROSOL

RAW MATERIALS	% By Weight
PRIMAPEL C-93 Polymer (25%)	12.0
Dipropylene Glycol Methyl Ether (Dowanol DPM)	0.1
Fluorad FC-129 (1%)	1.0
Nopcostat HS	0.5
Water	78.4
Isobutane	8.0
Perfume	As required
	100.0

Use Level: As Supplied

Properties:

Appearance	Milky Emulsion, separates
Stability	Stable
Freeze-Thaw Resistance	3 Cycles

SOURCE: Rohm and Haas Co.: Specialty Chemicals: Detergent
        Formulations for Industrial and Institutional Industry:
        Lit. Ref.: CS-502

LIQUID SPRAY VACUUM CARPET CLEANER

RAW MATERIALS	% By Weight
TRYLON 6735 Nonionic Surfactant	4.0
TRYCOL 5940 POE (6) Tridecyl Alcohol	1.0
TRYFAC 5569 Phosphate Ester	5.0
Triethanolamine (TEA)	2.5
Tetrapotassium pyrophosphate (TKPP)	5.0
Fragrance	as desired
Water	to 100

Blending Procedure:
   Add the water to the blending tank. While mixing, add the ingredients to the blending tank in the order listed. Stir until uniform.

Use Dilution:
   Dilute 1-2 ounces of the product per gallon of water before using.

SOURCE: Emery Chemicals: Specialty Chemicals Formulary:
       Formulation 2878-030

POWDERED SPRAY VACUUM CARPET CLEANER

RAW MATERIALS	% By Weight
Sodium Tripolyphosphate (STPP)	35.0
TRYLON 6735 Nonionic Surfactant	3.0
TRYCOL 5940 POE (6) Tridecyl Alcohol	1.0
Sodium metasilicate pentahydrate	3.0
Sodium sulfate (salt cake)	52.0
Sodium carbamate (soda ash)	6.0
Fragrance, optical brightener	as desired
	100.0

Blending Procedure:
   Premix the TRYLON 6735 and TRYCOL 5940 with any fragrance or brightener. Add them slowly to the STPP and mix thoroughly in a suitable powder blender. Add the remaining ingredients and mix until uniform.

Notes:
   1. Use Directions: Dilute 1/8 to 1/4 cup of formula per five gallons of water.
   2. Advise customers to test for colorfastness of carpet dyes before using the product.

SOURCE: Emery Chemicals: Specialty Chemicals Formulary:
       Formulation 2878-044

PREMIUM MEDIUM DUTY STEAM EXTRACTION CARPET CLEANER

RAW MATERIALS   % By Weight

Sodium Tripolyphosphate, Light Density   65.0
Sodium Metasilicate, Pentahydrate   29.5
PETRO 22 Powder   5.0
Optical Brightener   0.5

Blending Procedure: Blend ingredients in the order listed.
Use Dilutions: 1-2 oz/gal.

SOURCE: DeSoto, Inc.: Formulation 3/88: D-3082

PREMIUM HEAVY DUTY STEAM EXTRACTION CARPET CLEANER WITH BUTYL

RAW MATERIALS   % By Weight

Sodium Tripolyphosphate, Light Density   50.0
Butyl Cellosolve   4.0
Sodium Metasilicate, Pentahydrate   40.5
PETRO 22 Powder   5.0
Optical Brightener   0.5

Blending Procedure: Blend ingredients in the order listed.
Use Dilutions: 1-2 oz/gal.

SOURCE: DeSoto, Inc.: Formulation 3/88: D-3084

HEAVY DUTY STEAM EXTRACTION CARPET CLEANER

RAW MATERIALS   % By Weight

Sodium Carbonate (Soda Ash)   10.0
Trisodium Phosphate, Crystal   10.0
Sodium Metasilicate, Pentahydrate   20.0
Sodium Sulfate (Salt Cake)   56.0
PETRO 22 Powder   4.0

Blending Procedure: Blend ingredients in the order listed.

Use Dilution: 1-2 oz/gal.

SOURCE: DeSoto, Inc.: Formulation 10/87: D-3029

RUG CLEANER

RAW MATERIALS	% By Weight
DOWFAX 2A-1 surfactant (45% sol'n)	7
triethanolamine	10
DOWANOL PM glycol ether	4
potassium oleate	1
water	78
	low foam

SOURCE: Dow Chemical U.S.A.: The Glycol Ethers Handbook: I

RUG CLEANER

RAW MATERIALS	% By Weight
DOWANOL DPM glycol ether	5
trisodium phosphate	2
Orvus K liquid	7
water	86
	moderate foam

SOURCE: Dow Chemical U.S.A.: The Glycol Ethers Handbook: II

RUG CLEANER

RAW MATERIALS	% By Weight
DOWFAX 2A-1 surfactant	7
triethanolamine	10
DOWANOL DPM glycol ether	4
lauric diethanolamide	1
water	78
	high foam

SOURCE: Dow Chemical U.S.A.: The Glycol Ethers Handbook: III

AEROSOL RUG SHAMPOO WITH ANTICORROSIVE

RAW MATERIALS   % By Weight

Polymer Solution: SMA 1000A Powder   9.5
Ammonium Hydroxide 28%   6.1
Water   8.9
Sodium Lauryl Sulfate (Stepanol WAC)   0.5

Sodium Lauryl Sulfate (Stepanol WAC)   50.0
REWOCOR B3010   12.6
Water   11.4
Versene 100 (EDTA)   1.0
SMA 1000A, Ammonium Hydroxide Solution   25.0

Mixing Procedure

Dispense SMA powder in the water, add 2 grams of the SLS to
wet the powder. Gradually add the Ammonium Hydroxide with
stirring. Solution should become clear and hot as the ammonium
salt is formed. In a separate beaker mix in 50.0 g of SLS, 11.4
g of water, 1 gram of EDTA and REWOCOR B3010. Add phase I to
phase II using good agitation.

SOURCE: Sherex Chemical Co.: Industrial Formulation 27:1.4

RUG SHAMPOO WITH POLYMER

RAW MATERIALS   % By Weight

VARSULF SBL203   30.0
Sodium Lauryl Sulfate   33.0
Neo Cryl A550   20.0
Water   17.0

Mixing Procedure:
   Add Sodium Lauryl Sulfate to the water. Stir in VARSULF
SBL203 to the solution. Product should be clear. Then add the
Neo Cryl A550 polymer. The finished product should now be milky
but homogeneous.

SOURCE: Sherex Chemical Co.: Industrial Formulation 24:1.4

BASIC RUG SHAMPOO CONCENTRATE

RAW MATERIALS	% By Weight
HAMPOSYL L-30	30.0
Sodium lauryl sulfate, 30%	45.0
Formaldehyde, 37%	0.3
Water	24.7

Dilution for use: 1:40

SOURCE: W.R. Grace & Co.: Hampshire HAMPOSYL Surfactants:
           Suggested Formulation

RUG AND UPHOLSTERY SHAMPOO--LOW-FOAMING LIQUID

RAW MATERIALS	% By Weight
NEODOL 25-12	5.0
Sodium tripolyphosphate	2.0
Isopropyl alcohol	4.0
Water, dye, perfume	to 100%

Properties:
     Viscosity, 73F, cps   4
     Phase coalescence temp., F   >167
     pH   8.8
Recommended Dilution:
     For rugs and carpets: 1 part cleaner to 10 parts water
                       (12.8 oz/gal)
     For use on upholstery: 1 part cleaner to 20 parts water
                       (6.4 oz/gal).

SOURCE: Shell Chemical Co.: NEODOL Formulary: Suggested
           Formulation

RUG AND UPHOLSTERY SHAMPOO--HIGH-FOAMING LIQUID

RAW MATERIALS	% By Weight
NEODOL 25-3S (60%)	21.7
Water, dye, perfume	to 100%

Properties:
     Viscosity, 73F, cps   15
     Phase coalescence temp., F   >176
     pH   8.7
Recommended Dilution:
     1 part cleaner to 8 parts water (16 oz/gal).

SOURCE: Shell Chemical Co.: NEODOL Formulary: Suggested
           Formulation

RUG SHAMPOO

Amine soaps or synthetic detergents plus a solvent coupling agent act as excellent shampoos for cleaning rugs and upholstered furniture. The copious lather produced with water solutions of these shampoos washes easily, does not harm the fabric, and leaves the colors clear and bright. Use a sponge or soft cloth as an applicator. Press the excess liquid from the applicator, work up a lather on it, and rub the surface to be cleaned. Remove the soap by rubbing the cleaned surface with a clean cloth which has been wet with warm water. Wipe the surface with a clean dry cloth to remove the water. It is best to complete a small area at a time.

Amine Soap

	lbs.
Oleic Acid	23.5
Coconut-Oil Fatty Acids	17.5
99% Isopropanol	25.0
Triethanolamine	11.8
Monoethanolamine	5.6
SURFONIC N-95	4.2
Water	12.4

Mix the oleic acid, fatty acids, and isopropanol. Add the amines and SURFONIC N-95 and stir until thoroughly mixed. Then add the water, which will produce a clear liquid. Dilute the concentrated detergent with an equal volume of hot water before use.

The formula is based on a combining weight for coconut-oil fatty acids of 210. The proportion should be changed according to the combining weight of the particular fatty acid to be used. All the triethanolamine may be replaced by an additional 5.6 pounds of monoethanolamine to increase the detergency of the shampoo.

RUG SHAMPOO

RAW MATERIALS	Parts by Weight
Oleic Acid	28
Coconut Oil Fatty Acid	21
Isopropanol, Anhydrous	30
Triethanolamine	14
Monoethanolamine	6.8
SURFONIC N-95	5
Water	15

SOURCE: Texaco Chemical Co.: Suggested Formulations

RUG SHAMPOO

RAW MATERIALS	% By Weight	CAS Registry Number
Water	77.00	7732-18-5
Tetrasodium Phosphate	0.65	7720-88-5
Sodium Lauryl Sulfate @ 30%	9.20	151-21-3
ESI-TERGE B-15	3.00	61789-19-3
Butyl Cellosolve	1.00	111-76-2
Versene 100	0.15	64-02-8
	100.00	

Procedure:
   Add salts to water and dissolve.
   Add other ingredients in order mentioned.

Specifications:
   % Solids   10
   % Activity   11
   pH   10.5+-.5
   Viscosity   Water

SOURCE: Emulsion Systems Inc.: Technical Service Bulletin
        Code B15-4

RUG SHAMPOO
Liquid Type

RAW MATERIALS	% By Weight
Sodium lauryl sulfate	7.5
IGEPON TC-42	22.0
Isopropanol	3.0
Water	67.5
	100.0

Perfume and colorants added, as desired, replacing water.

Manufacturing Procedure:
   1. Dissolve sodium lauryl sulfate in water.
   2. Add IGEPON TC-42 and isopropanol.
   3. Filter product.

Physical Properties:
   pH (as is)   7.8
   pH (1%)   7.0
   Viscosity   10 cps
   Specific Gravity   1.01

SOURCE: GAF CORP.: Formulary: Prototype Formulation GAF 5701

RUG STEAM CLEANER

RAW MATERIALS   % By Weight

NEODOL 25-12	4.0
Sodium metasilicate, pentahydrate	8.0
Sodium xylene sulfonate (40%)	6.0
Water, dye, perfume	to 100%

Properties:
    Viscosity, 73F, cps   4
    Phase coalescence temp., F   >167
    pH   11.9

SOURCE: Shell Chemical Co.: NEODOL Formulary: Suggested
        Formulation

RUG STEAM CLEANER--NON PHOSPHATE POWDER

RAW MATERIALS   % By Weight

NEODOL 25-12	4.0
Sodium metasilicate, anhydrous	38.0
Sodium sulfate	25.0
Sodium carbonate	33.0

Blending Procedure:
    Mix solid builders and filler thoroughly.
    Add surfactant slowly while mixing, mix thoroughly.

SOURCE: Shell Chemical Co.: NEODOL Formulary: Suggested
        Formulation

RUG STEAM CLEANER--POWDER WITH PHOSPHATE

RAW MATERIALS   % By Weight

NEODOL 25-12	4.0
Sodium metasilicate, anhydrous basis	29.0
Sodium tripolyphosphate	15.0
Sodium sulfate	29.0
Sodium carbonate	23.0

Blending Procedure:
    Mix solid builders and filler thoroughly.
    Add surfactant slowly while mixing, mix thoroughly.

SOURCE: Shell Chemical Co.: NEODOL Formulary: Suggested
        Formulation

RUG AND UPHOLSTERY CLEANER--"DRY BRITTLE" LIQUID CLEANER--
HIGH QUALITY

RAW MATERIAL	% By Weight
NEODOL 25-12	13.0
NEODOL 25-3S (60%)	9.0
Tetrapotassium pyrophosphate (a)	7.0
Triton H-66	6.0
Water, dye, perfume	to 100%

Properties:
Viscosity, 73F, cps	13
Phase coalescence temp., F	150
pH	9.6

(a) Can use 1%w sodium metasilicate in place of the phosphate
    if a more alkaline product is desired.

SOURCE: Shell Chemical Co.: NEODOL Formulary: Suggested
        Formulation

RUG AND UPHOLSTERY CLEANER--"DRY BRITTLE" LIQUID CLEANER--
GOOD QUALITY

RAW MATERIALS	% By Weight
NEODOL 25-12	6.0
NEODOL 25-3S (60%)	3.0
Tetrapotassium pyrophosphate (a)	3.0
Water, dye, perfume	to 100%

Properties:
Viscosity, 73F, cps	13
Phase coalescence temp., F	>165
pH	10

(a) Can use 1%w sodium metasilicate in place of 1% of the
    phosphate if a more alkaline product is desired.

SOURCE: Shell Chemical Co.: NEODOL Formulary: Suggested
        Formulation

RUG AND UPHOLSTERY CLEANER--"DRY BRITTLE" LIQUID CLEANER--
ECONOMY

RAW MATERIALS	% By Weight
NEODOL 25-12	4.0
Tetrapotassium pyrophosphate (a)	8.0
Sodium xylene sulfonate (40%)	2.0
Water, dye, perfume	to 100%

Properties:
Viscosity, 73F, cps	3
Phase coalescence temp.,F	125
pH	10

(a) Can use 1%w sodium metasilicate in place of 1% of the phos-
    phate if a more alkaline product is desired.

SOURCE: Shell Chemical Co.: NEODOL Formulary: Suggested
        Formulary

LIQUID STEAM CLEANER FOR CARPETS
For Use In Hard Water Systems

RAW MATERIALS	% By Weight
Sodium metasilicate, anhydrous	7.5
Tetrapotassium pyrophosphate (60%)	12.5
GAFAC RA-600	4.0
IGEPAL CO-710	1.0
Water	75.0
	100.0

Perfume and colorants added, as desired, replacing water.

Manufacturing Procedure:
1. Dissolve sodium metasilicate, anhydrous in water.
2. Add GAFAC RA-600 and IGEPAL CO-710, mixing thoroughly
   after each addition.
3. Add tetrapotassium pyrophosphate.

Physical Properties:
pH (as is)	13.2
pH (1%)	11.0
Viscosity	10 cps
Specific Gravity	1.04

SOURCE: GAF Corp.: Formulary: Prototype Formulation GAF 5652

11. Wall and Hard Surface Cleaners

ALL PURPOSE HARD SURFACE CLEANING CONCENTRATES

RAW MATERIALS	% By Weight
MIRANOL C2M-SF CONC.	10.0
Actinol FA-2	6.0
Potassium Hydroxide, 45%	1.8
Ethylenediamine Tetra Acetic Acid	2.0
Pine Oil	1.0
Aqueous Ammonia, 28%	5.0
Sodium Tripolyphosphate	5.0
Water	69.2
MIRANOL C2M-SF CONC.	10.0
Dowanol EB	5.0
Sodium Carbonate	4.0
Ethylenediamine Tetra Acetic Acid	2.0
Sodium Xylene Sulfonate, 40%	4.0
Water	75.0

SOURCE: Miranol Inc.: MIRANOL Products for Household/Industrial
Applications: Suggested Formulations

HARD SURFACE CLEANER

RAW MATERIALS	% By Weight
DIACID H-240	3.5
Sodium silicate	8.0
Sodium carbonate	1.9
Neodol 15-S-9	3.1
Fatty acid (WESTVACO L-5)	1.0
Water	q.s.*

SOURCE: Westvaco Chemical Division: DIACID Surfactants:
Suggested Formulation

HARD SURFACE CLEANER

RAW MATERIALS	% By Weight
DIACID H-240	4.0
TSPP	4.2
Coco diethanolamide	7.8
Neodol 15-S-9	1.5
Tall oil fatty acid	2.0
Water	q.s.

* q.s.—quantity sufficient to make 100% total.

SOURCE: Westvaco Chemical Division: DIACID Surfactants:
Suggested Formulation

HARD SURFACE CLEANER

RAW MATERIALS	% By Weight
Water	88
Urea	2
Linear alkyl aryl sulfonate sodium salt (60%)	5
ICONOL DA-6 surfactant	2
Trilon B liquid [EDTA, tetrasodium salt (40%)]	2
Ammonium hydroxide (28%)	1

Formulation #3400

HARD SURFACE CLEANER

RAW MATERIALS	% By Weight
Water	85
Sodium xylene sulfonate (40%)	4
PLURAFAC B-25-5 surfactant	5
Sodium metasilicate pentahydrate	2
Tetrapotassium pyrophosphate	4

Formulation #3401

HARD SURFACE CLEANER

RAW MATERIALS	% By Weight
Water	91.5
Ethylene glycol butyl etner	5
PLURAFAC D-25 surfactant	1
Trilon B powder (EDTA, tetrasodium salt)	2.5

Spray type: use as is
Formulation #3403

HARD SURFACE CLEANER

RAW MATERIALS	% By Weight
Water	75.5
Ethylene glycol butyl ether	6
Sodium xylene sulfonate (40%)	3
Linear alkyl aryl sulfonate (60%)	3
PLURAFAC C-17 surfactant	5
PLURAFAC RA-40 surfactant	2.5
Sodium metasilicate pentahydrate	2
Tetrapotassium pyrophosphate	3

Suggested use concentration: 1-2 oz. per gallon of water and
   4-12 oz. per gallon of water for tough jobs
Formulation #3402

SOURCE: BASF Corp.: Cleaning Formulary: Suggested Formulations

HARD SURFACE CLEANER--ALL PURPOSE
CREAMY SCOURING CLEANSER

"Soft Scrub"* Type

High Quality

RAW MATERIALS	% By Weight
NEODOL 91-6**	4.5
NEODOL 91-2.5**	4.5
FADEA***	1.0
Calcium carbonate (100 mesh)****	40.0
Colloidal thickener*****	1.2
Organic gum******	0.4
Preservatives	0.1
Water, dye, perfume	to 100%

Good Quality

NEODOL 91-6**	6.3
NEODOL 91-2.5**	2.7
FADEA***	1.0
Calcium carbonate (100 mesh)****	40.0
Colloidal thickener*****	1.2
Organic gum******	0.4
Preservatives	0.1
Water, dye, perfume	to 100%

Blending Procedure:
   Blend thickener and organic gum and add to water slowly with
vigorous mixing until smooth. Add calcium carbonate slowly and
mix. Add NEODOLS and amide slowly and mix until smooth and
uniform. The final product is lotion-like. Can be made more
fluid by diluting with water.

         * Trademark of The Clorox Co.
        ** For household use, replace with NEODOL 23-6.5.
       *** Fatty acid diethanolamide, NINOL 2012EX, Stepan
           Chemical Co., or equivalent product.
      **** May substitute with KAOPOLITE, Georgia Kaolin or
           equivalent product.
     ***** VEEGUM, R.T. Vanderbilt Co., or equivalent product.
    ****** KELZAN (xanthan gum), Kelco Div., or equivalent product

SOURCE: Shell Chemical Co.: The NEODOL Formulary: Formulations

HARD SURFACE CLEANER: ALL-PURPOSE SPRAY

RAW MATERIALS	% By Weight
DIACID H-240	1.5
LAS	4.0
Neodol 23-6.5	2.0
Butyl cellosolve	5.0
Methoxy ethanol	2.0
Sodium carbonate	4.5
Water	q.s.*

HARD SURFACE CLEANER: LIQUID CONCENTRATE

RAW MATERIALS	% By Weight
DIACID H-240	1.4
Sodium metasilicate	1.0
Neodol 25-9	5.0
TKPP	10.0
Water	q.s.*

HARD SURFACE CLEANER: LIQUID DISINFECTANT

RAW MATERIALS	% By Weight
DIACID H-240	5.5
LAS	4.0
TKPP	10.0
O-Benzyl-p-chlorophenol	3.2
Isopropanol	2.0
Water	q.s.*

HARD SURFACE CLEANER: FLOOR CLEANER/WAX STRIPPER

RAW MATERIALS	% By Weight
DIACID H-240	2.3
Neodol 25-9	5.0
Trisodium phosphate	3.0
TKPP	5.0
Ammonium hydroxide	1.5
Water	q.s.*

*q.s.--quantity sufficient to make 100% total.

SOURCE: Westvaco Chemical Division: DIACID Surfactants:
Suggested Formulations

<u>HARD SURFACE CLEANER, LIQUID</u>
<u>All Surface Bathroom Type</u>

RAW MATERIALS   % By Weight

Water   70.5
IGEPAL CO-630   2.5
CHEELOX NTA-Na3   20.0
Sodium xylene sulfonate   5.5
Sodium hydroxide   1.5
                                                            100.0

Perfume and colorants added, as desired, replacing water.

Manufacturing Procedure:
   1. Dissolve IGEPAL CO-630 in water. Add CHEELOX NTA-Na3.
   2. Add sodium xylene sulfonate and sodium hydroxide, mixing
      well after each addition.

Physical Properties:
   pH (as is)   12.9
   pH (1%)   11.4
   Viscosity   10 cps
   Specific Gravity   1.05

SOURCE: GAF Corp.: Formulary: Prototype Formulation GAF 5157

<u>HARD SURFACE CLEANER</u>
All Surface Bathroom--Acid Type

RAW MATERIALS   % By Weight

M-PYROL   4.0
IGEPAL CO-630   1.0
Hydroxyethylcellulose   0.2
Phosphoric Acid   9.5
Water   85.3
                                                            100.0

Perfume and colorants added, as desired, replacing water.

Manufacturing Procedure:
   1. Disperse hydroxyethylcellulose.
   2. In a separate vessel, mix M-PYROL, IGEPAL CO-630 together.
      Add hydroxyethylcellulose/H20 mixture.
   3. Add acid, mix thoroughly.

Physical Properties:
   pH (as is)   2.0
   pH (1%)   2.4
   Viscosity   30 cps
   Specific Gravity   1.02

SOURCE: GAF Corp.: Formulary: Prototype Formulation GAF 5158

HARD SURFACE CLEANER
Degreasing Type

RAW MATERIALS	% By Weight
Water	85.2
Soda ash	2.0
Sodium metasilicate 5-H20	4.0
CHEELOX BF-13	2.6
Tetrapotassium pyrophosphate (60% active)	1.7
Sodium xylene sulfonate (40% active)	2.5
IGEPAL CA-630	2.0
	100.0

Manufacturing Procedure:
   Add ingredients in order, completely dissolving solids
   prior to addition of other components.

Physical Properties:
pH (as is)	12.8
pH (1%)	10.7
Viscosity	10 cps
Specific Gravity	1.02

SOURCE: GAF Corp.: Formulary: Prototype Formulation GAF 5161

HARD SURFACE CLEANER
All Purpose Type

RAW MATERIALS	% By Weight
Tetrapotassium pyrophosphate (60% active)	16.7
GAFAC RA-600	4.0
GAFAMIDE CDD-518	1.0
Alkylbenzene sulfonic acid	3.0
Water	75.3
	100.0

Manufacturing Procedure:
   1. Dissolve alkylbenzene sulfonic acid in total amount of
      water. Add GAFAC RA-600. Mix thoroughly.
   2. Add GAFAMIDE CDD-518 to main batch; add tetrapotassium
      pyrophosphate. Mix well.

Physical Properties:
pH (as is)	8.0
pH (1%)	8.0
Viscosity	80 cps
Specific Gravity	1.03

SOURCE: GAF Corp.: Formulary: Prototype Formulation GAF 5162

HARD SURFACE CLEANER
All Purpose Type

RAW MATERIALS	% By Weight
Tetrapotassium pyrophosphate (60% active)	12.5
GAFAC RA-600	4.0
IGEPAL CO-710	1.0
Sodium metasilicate, anhydrous	7.5
Water	75.0
	100.0

Manufacturing Procedure
1. Dissolve IGEPAL CO-710 in water. Add sodium metasilicate, anhydrous, to surfactant/water mixture. Mix thoroughly.
2. Add GAFAC RA-600 and tetrapotassium pyrophosphate, mixing well after each addition.

Physical Properties:
pH (as is)	13.4
pH (1%)	11.2
Viscosity	10 cps
Specific Gravity	1.05

SOURCE: GAF Corp.: Formulary: Prototype Formulation GAF 5163

HARD SURFACE CLEANER
All Purpose Type

RAW MATERIALS	% By Weight
IGEPAL CA-630	2.0
GAFAC RA-600	4.0
Sodium metasilicate, anhydrous	3.5
Tetrapotassium pyrophosphate (60% active)	10.8
Water	79.7
	100.0

Manufacturing Procedure:
1. Dissolve IGEPAL CA-630 in water. Add sodium metasilicate, anhydrous, to IGEPAL CA-630 and water mixture. Agitate until mixture is homogeneous.
2. Add GAFAC RA-600 and tetrapotassium pyrophosphate, mixing thoroughly after each addition.

Physical Properties:
pH (as is)	13.1
pH (1%)	10.7
Viscosity	10 cps
Specific Gravity	1.03

SOURCE: GAF Corp.: Formulary: Prototype Formulation GAF 5164

HARD SURFACE CLEANER
All Purpose Type

RAW MATERIALS	% By Weight
M-PYROL	5.0
Trisodium pnosphate	2.0
Sodium metasilicate 5-H2O	2.0
CHEELOX BF-78	2.2
EMULPHOGENE BC-720	7.0
Sodium xylene sulfonate (40%)	1.0
Water	80.8
	100.0

Manufacturing Procedure:
1. Add sodium xylene sulfonate to total amount of water.
   Add EMULPHOGENE BC-720 to solution.
2. Add sodium metasilicate 5-H2O, trisodium phosphate and
   CHEELOX BF-78, mixing thoroughly after each addition.
3. Add M-PYROL. Mix well.

Physical Properties:
pH (as is)	12.6
pH (1%)	10.3
Viscosity	10 cps
Specific Gravity	1.01

SOURCE: GAF Corp.: Formulary: Prototype Formulation GAF 5165

HARD SURFACE CLEANER

RAW MATERIALS	% By Weight
Water	75.0
Sodium Carbonate	4.0
VARION AMKSF 40	10.0
Dowanol DPM Solvent	5.0
Versene 100	2.0
Whitconate SXS 40	4.0

Mixing Procedure:
   Add ingredients as listed.

SOURCE: Sherex: Industrial Formulation 34:1.3

HARD SURFACE CLEANER
General Purpose Type

RAW MATERIALS	% By Weight
Alkylbenzene sulfonic acid	1.0
M-PYROL	3.0
Isopropanol	1.0
IGEPAL CO-710	2.4
CHEELOX BF-13	2.6
Water	90.0
	100.0

Perfumes and colorants added, as desired, replacing water.

Manufacturing Procedure:
1. The ingredients are added individually, in order, mixing well after each addition.

Physical Properties:

pH (as is)	9.8
pH (1%)	9.6
Viscosity	10 cps
Specific Gravity	1.01

SOURCE: GAF Corp.: Formulary: Prototype Formulation GAF 5159

HARD SURFACE CLEANER
Heavy Duty Type

RAW MATERIALS	% By Weight
M-PYROL	5.0
GAFAC RA-600	4.0
IGEPAL CO-710	1.0
Carboxymethylcellulose	2.0
Water	74.3
Tetrapotassium pyrophosphate (60% active)	11.7
Potassium hydroxide (50% active)	2.0
	100.0

Manufacturing Procedure:
1. Disperse carboxymethylcellulose completely in water in a separate vessel.
2. Add M-PYROL, GAFAC RA-600 and IGEPAL CO-710 together.
3. Add carboxymethylcellulose/water mixture to main batch.
4. Add tetrapotassium pyrophosphate, potassium hydroxide.
5. Filter product.

Physical Properties:

pH (as is)	11.4
pH (1%)	9.6
Viscosity	60 cps
Specific Gravity	.98

SOURCE: GAF Corp.: Formulary: Prototype Formulation GAF 5160

HARD SURFACE CLEANER
Pine Oil Type--Economical

RAW MATERIALS   % By Weight

Pine Oil   5.0
EMULPHOGENE BC-840   10.0
Water   85.0
                                                           100.0

Manufacturing Procedure:
1. Add EMULPHOGENE BC-840 to pine oil. Mix well. Heat to
   ⁻35-40C to obtain homogeneous solution.
2. Add water and mix thoroughly.

Physical Properties:
    pH (as is)   2.8
    pH (1%)   4.7
    Viscosity   10 cps
    Specific Gravity   1.00

SOURCE: GAF Corp.: Formulary: Prototype Formulation GAF 5155

HARD SURFACE CLEANER
Solvent Type

RAW MATERIALS   % By Weight

Odorless mineral spirits   25.0
Pine oil   10.0
IGEPAL CO-630   5.0
Tall oil fatty acid   10.0
Water   45.0
Tetrapotassium pyrophosphate (60% active)   5.0
Potassium hydroxide (50% active, until product clears)   q.s.
                                                           100.0

Manufacturing Procedure:
1. Add components in order listed above, mixing thoroughly
   after each addition.
2. Add potassium hydroxide dropwise, until product is a
   clear yellow color.

Physical Properties:
    pH (as is)   9.1
    pH (1%)   9.5
    Viscosity   130 cps
    Specific Gravity   .99

SOURCE: GAF Corp.: Formulary: Prototype Formulation GAF 5156

HARD SURFACE CLEANERS--ALL PURPOSE
ALL PURPOSE LIQUID CONCENTRATE

"Janitor in a Drum"* Type

RAW MATERIALS	% By Weight
NEODOL 23-6.5	5.0
NEODOL 25-3	2.5
Butyl OXITOL	6.0
Pine oil	0.25
Tetrapotassium pyrophosphate	3.0
Sodium metasilicate, pentahydrate	2.0
Sodium xylene sulfonate (40%)	1.0
Water, dye, perfume	to 100%

Properties:
<pre>
    Viscosity, 73F, cps   5
    Phase coalescence temp., F   104
    pH   13
</pre>

Use Concentration:
    2-4 oz/gal.

ALL PURPOSE SPRAY

"Fantastic"* Type

RAW MATERIALS	% By Weight
NEODOL 23-6.5 or	
NEODOL 25-3S (60%)	1.7
Cocodiethanolamide	0.5
Trisodium phosphate, anhydrous basis	1.0
Sodium metasilicate, pentahydrate	1.7
Butyl OXITOL	3.5
Water, dye, perfume	to 100%

Properties:
<pre>
    Viscosity, 73F, cps   7
    Phase coalescence temp., F   140
    pH   12.4
</pre>

   * Trademark of Dow Consumer Products, Inc

SOURCE: Shell Chemical Co.: The NEODOL Formulary: Formulations

HARD SURFACE CLEANERS--ALL PURPOSE
NON-PHOSPHATE PREMIUM QUALITY LIQUID CONCENTRATES

For Hard Water

RAW MATERIALS	% By Weight
NEODOL 91-6	7.0
Sodium metasilicate, pentahydrate	14.6
EDTA	12.6
Sodium xylene sulfonate (40%)	30.0
Water, dye, perfume	to 100%

Properties:
Viscosity, 73F, cps	13
Phase coalescence temp., F	>140
pH	12.9

Recommended Dilutions:
Heavy-duty use: 4 oz/gal.
Regular-duty use: 2 oz/gal.

For Medium-Hardness Water

RAW MATERIALS	% By Weight
NEODOL 91-6	8.0
Sodium metasilicate, pentahydrate	11.1
EDTA	9.6
Sodium xylene sulfonate (40%)	30.0
Water, dye, perfume	to 100%

Properties:
Viscosity, 73F, cps	11
Phase coalescence temp., F	>140
pH	12.7

Recommended Dilutions:
Heavy-duty use: 4 oz/gal.
Regular-duty use: 2 oz/gal.

Blending Procedure:
Dissolve the surfactant and hydrotrope in water. Add the builders with stirring at a rate to promote solution. For long term stability reasons the formulated concentrates should not be stored in glass containers.

SOURCE: Shell Chemical Co.: The NEODOL Formulary: Suggested Formulations

HARD SURFACE CLEANERS--ALL PURPOSE
NON-PHOSPHATE PREMIUM QUALITY LIQUID CONCENTRATES
For Heavy Oil Removal

RAW MATERIALS	% By Weight
NEODOL 91-6	3.0
NEODOL 91-2.5	3.0
Sodium metasilicate, pentahydrate	12.5
NTA	4.8
TRITON H-66	8.0
Water, dye, perfume	to 100%

Properties:

Viscosity, 73F, cps	13
Phase coalescence temp., F	>140
pH	13.2

Recommended Dilutions:
    Heavy-duty use: 12.8 oz/gal (1/10)
    Regular-duty use: 4 oz/gal (1/32)
    Light-duty use: 2 oz/gal (1/64)

HARD SURFACE CLEANER LIQUID CONCENTRATES
High Quality with Phosphate

RAW MATERIALS	% By Weight
NEODOL 91-6	2.0
NEODOL 91-2.5	2.0
Sodium metasilicate, pentahydrate	8.3
Trisodium phosphate, anhydrous basis	3.2
Triton H-66	7.0
Water, dye, perfume	to 100%

Properties:

Viscosity, 73F, cps	8
Phase coalescence temp., F	150
pH	13.2

Recommended Dilutions:
    Heavy-duty use: 12.8 oz/gal (1/10)
    Regular-duty use: 8 oz/gal (1/16)
    Light-duty use: 4 oz/gal (1/32)

Blending Procedure:
    Dissolve the surfactant and hydrotrope in water. Add the
builders with stirring at a rate to promote solution.

    * May substitute 12% sodium xylene sulfonate (40%)

SOURCE: Shell Chemical Co.: The NEODOL Formulary: Suggeested
        Formulations

HARD SURFACE CLEANERS--ALL PURPOSE
HARD SURFACE CLEANER LIQUID CONCENTRATES
High Quality Non-Phosphate

RAW MATERIALS	% By Weight
NEODOL 91-6	2.0
NEODOL 91-2.5	2.0
Sodium metasilicate, pentahydrate	8.3
NTA	3.2
Triton H-66*	7.0
Water, dye, perfume	to 100%

Properties:

Viscosity, 73F, cps	8
Phase coalescence temp., F	135
pH	13.2

Regular Quality with Phosphate

RAW MATERIALS	% By Weight
NEODOL 91-6	2.8
NEODOL 91-2.5	1.2
Sodium metasilicate, pentahydrate	11.1
Trisodium phosphate, anhydrous basis	1.6
Triton H-66**	6.0
Water, dye, perfume	to 100%

Properties:

Viscosity, 73F, cps	5
Phase coalescence temp., F	135
pH	13.2

Regular Quality Non-Phosphate

RAW MATERIALS	% By Weight
NEODOL 91-6	2.8
NEODOL 91-2.5	1.2
Sodium metasilicate, pentahydrate	11.1
NTA	1.6
Triton H-66**	6.0
Water, dye, perfume	to 100%

Properties:

Viscosity, 73F, cps	8
Phase coalescence temp., F	135
pH	13.2

  * May substitute with 12% sodium xylene sulfonate (40%)
 ** May substitute with 10% sodium xylene sulfonate (40%)

Recommended Dilutions:
    Heavy-duty use: 12.8 oz/gal (1/10)
    Regular-duty use: 8 oz/gal (1/16)
    Light-duty use: 4 oz/gal (1/32)

SOURCE: Shell Chemical Co.: The NEODOL Formulary: Formulations

HARD SURFACE CLEANERS--ALL PURPOSE

Economy with Phosphate

RAW MATERIALS	% By Weight
NEODOL 91-6*	1.8
NEODOL 91-2.5*	0.8
Sodium metasilicate, pentahydrate	5.6
Trisodium phosphate, anhydrous basis	1.6
Sodium carbonate	3.0
Triton H-66**	6.0
Water, dye, perfume	to 100%

Properties:
Viscosity, 73F, cps   10
Phase coalescence temp., F   160
pH   13.2

Economy Non-Phosphate

RAW MATERIALS	% By Weight
NEODOL 91-6*	1.8
NEODOL 91-2.5*	0.8
Sodium metasilicate, pentahydrate	5.6
NTA	1.6
Sodium carbonate	3.0
Triton H-66**	6.0
Water, dye, perfume	to 100%

Properties:
Viscosity, 73F, cps   8
Phase coalescence temp., F   160
pH   13.2

 * May substitute with NEODOL 23-6.5
** May substitute with 10% sodium xylene sulfonate (40%).

Blending Procedure:
   Dissolve the surfactant and hydrotrope in water. Add the
builders with stirring at a rate to promote solution. For
long term stability reasons the formulated concentrates should
not be stored in glass containers.

Recommended Dilutions:
   Heavy-duty use: 12.8 oz/gal (1/10)
   Regular-duty use: 8 oz/gal (1/16)
   Light-duty use: 4 oz/gal (1/32)

SOURCE: Shell Chemical Co.: The NEODOL Formulary: Formulations

HARD SURFACE CLEANERS--ALL PURPOSE
PINE OIL CLEANERS

High Quality

RAW MATERIALS	% By Weight
Pine oil	20.0
NEODOL 91-8	4.7
C12 LAS (60%)*	7.8
Isopropyl alcohol	11.0
Triethanolamine	4.7
Water, dye	to 100%

Properties:
Viscosity, 73F, cps	16
Phase coalescence temp., F	>158
pH	10.7
Cloudiness on mixing with water	very high

Good Quality

RAW MATERIALS	% By Weight
Pine oil	10.0
NEODOL 91-8	4.7
C12 LAS (60%)*	7.8
Isopropyl alcohol	8.0
Triethanolamine	2.0
Water, dye	to 100%

Properties:
Viscosity, 73F, cps	15
Phase coalescence temp., F	>158
pH	10.3
Cloudiness on mixing with water	high to average

Blending Procedure:
   Simple mixing, If using NEODOL 25-3S, add it last, slowly
with good stirring.

   * Witconate 1260, Witco Chemical Co., or equivalent product.
   May substitute with equal amounts of NEODOL 25-3S (60%)
   plus 4.6% (for the high quality formula) or 2.0% (for the
   good quality formula) sodium xylene sulfonate (40%).

SOURCE: Shell Chemical Co.: The Neodol Formulary: Suggested
        Formulations

HARD SURFACE CLEANERS--ALL PURPOSE

READY-TO-USE LIQUID

"Mr. Clean"* Type

RAW MATERIALS	% By Weight
NEODOL 23-6.5	4.8
C12 LAS (60%)	3.3
Sodium carbonate	4.0
EDTA**	6.0
Sodium xylene sulfonate (40%)	6.0
Water, dye, perfume	to 100%

Properties:

Viscosity, 73F, cps	23
Phase coalescence temp., F	>176
pH	11.3

Use Concentration: 2 oz/gal.

SPRAY AND WIPE LIQUID

RAW MATERIALS	% By Weight
NEODOL 91-6	3.0
Butyl OXITOL	3.0
Sodium tripolyphosphate	2.0
Sodium metasilicate, pentahydrate	2.0
EDTA**	0.5
Sodium xylene sulfonate (40%)	4.0
Water, dye, perfume	to 100%

Properties:

Viscosity, 73F, cps	5
Phase coalescence temp., F	142
pH	13

Blending Procedure:
    Dissolve phosphate and silicate in warm water. Add EDTA, NEODOL 91-6, sodium xylene sulfonate and Butyl OXITOL with agitation until homogeneous.

    * Trademark of Proctor and Gamble
   ** Ethylenediamine tetraacetic acid, tetrasodium salt (100% basis). Can be replaced with nitrilotriacetic acid, trisodium salt or sodium citrate. If replaced with sodium citrate, must increase sodium xylene sulfonate to 10% for good temperature stability.

SOURCE: Snell Chemical Co.: The NEODOL Formulary: Formulations

HARD SURFACE SANITIZER AND CLEANER DISINFECTANT

RAW MATERIALS	% By Weight
BIOPAL NR-20	8.8
IGEPAL CO-660	15.0
Phosphoric acid (85%)	7.0
Water	69.2
	100.0

Manufacturing Procedure:
1. Dissolve phosphoric acid and IGEPAL CO-660 in water until clear.
2. Add BIOPAL NR-20 to mixture using moderate agitation. Continue agitation until uniform.

Physical Properties:
pH (as is)	1.6
pH (1% solution)	2.7
Viscosity	510 cps
Specific Gravity	1.02

Note:
Each formulator is responsible for obtaining EPA registration for its end use product. GAF has made this process easier for you by registering this formulation under the names BIOPAL NR-I (EPA Reg. No. 1529-24) and BIOPAL NR-II (EPA Reg. No. 1529-23).

SOURCE: GAF Corp.: Formulary: Prototype Formulation GAF 5978

HARD SURFACE CLEANER/FABRIC PRETREAT
Grease Remover Type

RAW MATERIALS	% By Weight
Sodium lauryl sulfate	2.0
IGEPAL CO-630	12.0
EMULPHOGENE DA-630	6.0
Water	80.0
	100.0

Perfume and colorants added, as desired, replacing water.

Manufacturing Procedure:
1. Dissolve sodium lauryl sulfate in water with small amount of heat, 30-35C, if needed.
2. Add IGEPAL CO-630 and EMULPHOGENE DA-630 individually, mixing well after each addition.

Physical Properties:
pH (as is)	6.7
pH (1%)	6.1
Viscosity	40 cps
Specific Gravity	1.01

SOURCE: GAF Corp.: Formulary: Prototype Formulation GAF 5103

HARD SURFACE SPRAY CLEANER

RAW MATERIALS	% By Weight
TRITON X-102 Surfactant	1.0
Tetrapotassium Pyrophosphate (TKPP)	2.5
Dipropylene Glycol Methyl Ether (Dowanol DPM)	5.0
Water	91.5
	100.0

Directions for Use: Spray on as prepared. Wipe off.

SOURCE: Rohm and Haas Co.: Specialty Chemicals: Detergent
        Formulations for Industrial and Institutional Industry:
        Lit. Ref: CS-407

HARD SURFACE AND FLOOR CLEANER

RAW MATERIALS	% By Weight
I:	
Trisodium phosphate	3
Sodium tripolyphosphate or	
tetrasodium pyrophosphate	2
STEPANATE X (40% active)	6
NINOL 11-CM	8
Water, dye, perfume	q.s.
II:	
STEPANATE X (40% active)	5
Sodium metasilicate, anhydrous	3
Na3NTA	2.5
NINOL 11-CM	8
Water, dye, perfume	q.s.

Mixing Procedure:
    Dissolve builders in water, add STEPANATE X, and then NINOL
11-CM.

Properties:	I.	II.
Appearance	clear,light yellow liquids	
Viscosity @ 25C, cps	110	100
pH, as is	11.2	11.5
Active, %	15.4	15.9
Freeze/thaw stability	passes 3 cycles	

Use Instructions:
    These concentrates can be used in a dilution of 2 to 4
oz/gal.
Performance:
    Excellent formulations for use in manual or completely auto-
mated scrub machines because of moderate, fast-breaking foam
which does not interfere with machine operation.

SOURCE: Stepan Co.: Formulation No. 82

LIQUID HARD SURFACE CLEANER
High Sudsing--Unbuilt

RAW MATERIALS	% By Weight
Water	79.1
Alkylbenzene sulfonic acid (90% active)	4.7
Sodium hydroxide (50% active)	1.2
Sodium xylene sulfonate (40% active)	3.0
Sodium silicate (2.4 ratio solids)	5.5
IGEPAL CO-710	1.0
GAFAMIDE CDD-518	2.0
M-PYROL	3.5
	100.0

Perfume, colorant and opacifier added, as desired, replacing water.

Manufacturing Procedure:
   Add components in order listed. (Note: To avoid formation of silica floc, the product should be neutral or slightly alkaline prior to sodium silicate addition.)

Physical Properties:
pH (as is)	12.5
pH (1%)	10.8
Viscosity	10 cps
Specific Gravity	1.06

SOURCE: GAF Corp.: Formulary: Prototype Formulation GAF 5153

LIQUID HARD SURFACE CLEANER
Dilutable, Phosphate Built

RAW MATERIALS	% By Weight
Water	52.3
ALIPAL CD-128	8.3
M-PYROL	3.0
Ammonia (28% active)	3.4
Tetrapotassium pyrophosphate (60% active)	33.0
	100.0

Perfume and colorants added, as desired, replacing water.

Manufacturing Procedure:
   1. Add components in the order listed. (Minimize aeration to avoid foaming.)
   2. Tetrapotassium pyrophosphate may be added as 100% solids; balance the water accordingly.

Physical Properties:
pH (as is)	11.5
pH (1%)	9.6
Viscosity	10 cps
Specific Gravity	1.19

SOURCE: GAF Corp.: Formulary: Prototype Formulation GAF 5151

LIQUID HARD SURFACE CLEANER CONCENTRATES

RAW MATERIALS	% By Weight
MIRANOL J2M CONC.	6.7
Potassium Hydroxide, 45%	8.6
Kasil #1	25.7
Tetrapotassium Pyrophosphate	13.3
Water	45.7
MIRANOL J2M CONC.	5.7
Potassium Hydroxide, 45%	6.7
Kasil #1	19.2
Tetrapotassium Pyrophosphate	10.0
Water	58.4

SOURCE: Miranol Inc.: MIRANOL Products for Household/Industrial
Applications: Suggested Formulations

WALL CLEANER

RAW MATERIALS	% By Weight
DIACID H-240	4.6
Alkylolamide	5.0
Pluronic L-61	1.0
TKPP	10.0
Butyl cellosolve	2.0
Water	q.s.*

* q.s.--quantity sufficient to make 100% total.

SOURCE: Westvaco Chemical Division: DIACID Surfcactants:
Suggested Formulation

WALL CLEANER, LIGHT DUTY

RAW MATERIALS	% By Weight
TRITON X-100 Surfactant	5.0
Trisodium Phosphate (TSP)	2.0
Sodium Metasilicate Pentahydrate	2.0
Water	91.0
	100.0

Use Dilution: 1 part in 4 parts water.
Lit. Ref: CS-427

WALL CLEANER, SPRAY

RAW MATERIALS	% By Weight
TRITON X-114 Surfactant	2.5
Tetrasodium Ethylenediaminetetraacetate (Versene 100)	2.5
Dipropylene Glycol Methyl Ether (Dowanol DPM)	3.0
Isopropyl Alcohol	1.0
Water	91.0
Colorants, Perfume	Optional

Directions For Use: Spray as prepared
Lit. Ref: CS-409
SOURCE: Rohm and Haas Co.: Suggested Formulations

12. Window and Glass Cleaners

ACID GLASS CLEANER B-1

RAW MATERIALS   % By Weight

Ethylene Glycol Monobutylether	10.0
AVANEL S-70	0.3
Deionized Water	q.s. to 100.0
Color	As desired
Acetic Acid	to pH 3.5-4.0

This formula was developed as a glass cleaner to remove
common soils and hard water spots. The ratio of surfactants to
solvents gives uniform cleaning without streaking even under
hot summer conditions. The formulation gives superior performance because of the ability of the anionic AVANEL S
surfactants to maintain their surface activity under acid
conditions and because of the high grease solubilization power
of the AVANEL S-70. This formulation is also effective on surfaces other than glass normally found in the home such as
counter tops and painted surfaces.

Preparation Method
    Charge the Ethylene Glycol Monobutylether, AVANEL S-70 and
most of the water. Dissolve the dye in the portion of the
water charge held out and add to the product. Adjust pH to
3.5 to 4.0 with acetic acid.

SOURCE: Mazer Chemicals, Inc.: AVANEL S Formula

MULTI-FEATURE GLASS CLEANER

RAW MATERIALS   % By Weight

Isopropyl Alconol	10-15
Water	85
MAZER MACOL 19	1.75
MAZER MASIL 1066C	.25

SOURCE: Mazer Chemicals, Inc.: Household/Industrial T-20B:
        Formula 8

AMMONIACAL "WINDEX" (b) TYPE

RAW MATERIALS	% By Weight
NEODOL 25-3A(60%)	0.15
Isopropyl alcohol	5.0
Ammonia, conc.	0.15
Water, dye	to 100%

Properties:

Viscosity, 73F, cps	5
Phase coalescence temp., F	>176
pH	9.8

(b) Trademark of the Drackett Co.

SOURCE: Shell Chemical Co. NEODOL Formulary: Suggested Formula

WINDOW/GLASS CLEANER (GOOD QUALITY)

RAW MATERIALS	% By Weight
NEODOL 25-3S (60%)	0.15
Tetrapotassium pyrophosphate	0.02
Butyl OXITOL	0.10
Water, dye	to 100%

Properties:

Viscosity, 73F, cps	4.5
Phase coalescence, temp., F	>176
pH	8.9

SOURCE: Shell Chemical Co.: NEODOL Formulary: Suggested Formula

<u>GLASS CLEANER</u>

RAW MATERIALS   % By Weight

Water   91.6
Ethylene glycol butyl ether   3.5
Iconol TDA-8 surfactant   1.7
Coconut diethanolamide   0.5
Sodium metasilicate pentahydrate   1.7
Tetrapotassium pyrophosphate   1

Use as is

SOURCE: BASF Corp.: Cleaning Formulary: Formulation #3451

<u>GLASS CLEANER(ALL PURPOSE)</u>

RAW MATERIALS   % By Weight

DOWANOL PM glycol ether   8.0
ammonium hydroxide (28%)   1.5
PLURONIC F108 surfactant   0.1
water   90.4

   This formulation is particularly suited for packaging in a
spray bottle.
   The concentration of DOWANOL PM can be adjusted to control
the evaporation rate and appearance on the glass.

SOURCE: Dow Chemical U.S.A.: The Glycol Ethers Handbook:
        Formulation III.

GLASS CLEANER(ALL PURPOSE)

RAW MATERIALS   % By Weight

DOWANOL PM or
  DOWANOL DPM glycol ether   5
propylene glycol   5
isopropanol   35
water   55

    This formulation is particularly suited for packaging in a spray bottle.
    The concentration of DOWANOL PM, DOWANOL DPM and isopropanol can be adjusted to control the evaporation rate and appearance on the glass.

SOURCE: Dow Chemical U.S.A.: The Glycol Ethers Handbook:
       Formulation I.

GLASS CLEANER(ALL PURPOSE)

RAW MATERIALS   % By Weight

DOWANOL DPM glycol ether   4.0
isopropanol   4.0
ammonium hydroxide (28%)   1.0
PLURONIC F108 surfactant   0.1
water   90.9

    This formulation is particularly suited for packaging in a spray bottle.
    The concentration of DOWANOL DPM and isopropanol can be adjusted to control the eveporation rate and appearance on the glass.

SOURCE: Dow Chemical U.S.A.: The Glycol Ethers Handbook:
       Formulation II.

GLASS CLEANER, READY TO USE

RAW MATERIALS   % By Weight

Water, D.I.   84.9
PETRO BAF Liquid   0.1
Isopropyl Alcohol   10.0
Butyl Cellosolve   5.0
Dye   q.s.

Blending Procedure:
    Blend ingredients in the order listed.

SOURCE: DeSoto, Inc.: Formulation 3/88: K-3079

GLASS CLEANER

RAW MATERIALS   % By Weight

Water   89.1
MAZER MASIL 1066C   .2
MAZER MAZAWET DF   .2
MAZER MACOL 212   .5
Isopropyl Alcohol   10.0

SOURCE: Mazer Chemicals, Inc.: Household/Industrial T-20B:
        Formulation 9

GLASS CLEANER
General Purpose Type

Raw Materials	% By Weight
M-PYROL	4.0
Isopropanol	4.0
Ammonium Hydroxide (28% active)	1.0
ANTAROX BL-240	0.1
Water	90.9

Manufacturing Procedure:
1. Dissolve ANTAROX BL-240 in water. Add M-PYROL. Mix thoroughly.
2. Add isopropanol and ammonium hydroxide to main batch.

Physical Properties:
pH (as is)	10.4
pH (1%)	8.9
Viscosity	10 cps
Specific Gravity	1.00

SOURCE: GAF CORP.: Formulary: Prototype Formulation GAF 5402

GLASS CLEANER CONCENTRATE

RAW MATERIALS	% By Weight
Water, D.I.	61.5
PETRO BA or BAF Powder	2.5
Tetrapotassium Pyrophosphate	1.0
Butyl Cellosolve	5.0
Sodium Gluconate (Tech.)	1.0
Ammonium Hydroxide	5.0
Isopropyl Alcohol (99%)	24.0

Blending Procedure:
Blend ingredients with mild agitation in the order listed.

Use Dilutions:
For general or light glass cleaning, mix 16:1. For heavy buildup, mix 8:1.

SOURCE: DeSoto, Inc.: Formulation 3/88: K-3060

INDUSTRIAL GLASS CLEANER

RAW MATERIALS   % By Weight

Water	75
MAZER MACOL NP 9.5	3
Sodium Dihydrogen Phosphate	5
MAZER MACOL 212	5
MAZER MAZON DDBSA	3
MAZER MAZAWET DF	4
Isopropyl Alcohol	3

Procedure:
    Mix in order of listing

SOURCE: Mazer Chemicals, Inc.: Household/Industrial T-20B: 22

LIQUID GLASS CLEANER

RAW MATERIALS   % By Weight

Isopropyl Alcohol (Anhydrous)	5.00
Butyl Cellosolve Glycol Ether	3.00
TRITON X-100 Surfactant	0.05
Water	91.95

Add dye or perfume, as required.
Use as prepared.

Lit. Ref.: CS-427

SOURCE: Rohm and Haas Co.: Specialty Chemicals: Detergent
        Formulations for Industrial and Institutional Industry

WINDOW/GLASS CLEANER(GOOD QUALITY WITH ALCOHOL)

RAW MATERIALS	% By Weight
NEODOL 23-6.5	0.1
Isopropyl alcohol	15.0
Water, dye	to 100%

Properties:
Viscosity, 73F, cps	6
Phase coalescence temp., F	>176
pH	7.5

SOURCE: Shell Chemical Co.: NEODOL Formulary: Suggested
        Formulation

WINDOW AND GLASS CLEANER

RAW MATERIALS	% By Weight
Isopropanol	35.0
Propylene Glycol Monomethyl Ether	7.5
SURFONIC N-95	0.5
Water	57.0

Mix other ingredients thoroughly before adding the SURFONIC
N-95.

SOURCE: Texaco Chemical Co.: Formulation PEG12/W2

WINDOW AND GLASS CLEANER

RAW MATERIALS	% By Weight
Isopropanol	35.0
Propylene glycol monomethyl ether	7.5
SURFONIC N-95	0.5
Water	57.0

SOURCE: Texaco Chemical Co.: SURFONIC N-Series Surface-Active
Agents: Suggested Formulation

WINDOW CLEANER

RAW MATERIALS	% By Weight
MIRANOL JEM CONC.	0.3
Isopropyl Alcohol	8.0
Dowanol EB	1.0
Water	90.7

SOURCE: Miranol Inc.: MIRANOL Products for Household/Industrial
Applications: Suggested Formulation

WINDOW CLEANER
Pump Dispensing Type

RAW MATERIALS	% By Weight
Isopropanol	10.0
M-PYROL	5.0
ANTAROX BL-225	0.1
Ammonium Hydroxide (30% active)	1.0
Water	83.9
	100.0

Manufacturing Procedure:
1. Add components individually, in above order, mixing well
   after each addition.

Physical Properties:
pH (as is)	9.3
pH (1%)	8.4
Viscosity	10 cps
Specific Gravity	1.01

SOURCE: GAF Corp.: Formulary: Prototype Formulation GAF 5401

WINDOW CLEANER (2878-043)

RAW MATERIALS   % By Weight

Isopropyl alcohol   10.0
Ethylene glycol n-butyl ether   2.0
TRYCOL 6964 POE (9) Nonylphenol   0.1
EMERSAL 6400 Sodium Lauryl Sulfate   0.3
26 Be aqueous ammonia   0.2
Dye, fragrance, etc.   as desired
Water   to 100

Blending Procedure:
    Add the water to the blending tank. While mixing, add the
remaining ingredients in the order listed.

Note: Nonvolatile ingredients, such as POE (9) nonylphenol,
glycol ether and sodium lauryl sulfate should not be increased
above the levels recommended here. Although they aid in cleaning,
higher levels result in streaking if they are not completely
buffed off the windows.

SOURCE: Emery Chemicals: Specialty Chemicals Formulary:
        Formulation 2878-043

WINDOW CLEANER
AEROSOL (FOAM-TYPE)

RAW MATERIALS   % By Weight

DOWANOL DPM glycol ether   5.00
DOWANOL PM glycol ether   5.00
Pluronic F108 surfactant   0.10
colloidal silica (Ludox Tech.)   4.00
DUPONOL C surfactant   0.05
water   85.85

    To each 100 parts of the above formulation, add 0.3 parts
sodium benzoate.

Packaging Instructions:
    Above formulation   95.0 wt. %
    Propellant A-31   5.0

Can: Standard tin or lacquer-lined
Valve: Buna gasket for water-based products with mechanical
       breakup button.

SOURCE: Dow Chemical Co.: The Glycol Ethers Handbook: Suggested
        Formulation

WINDOW CLEANER
HIGH ALCOHOL TYPE

RAW MATERIALS   % By Weight

Isopropanol   47.5
IGEPAL CA-620   2.5
Water   50.0
                                                                     100.0

Manufacturing Procedure
    Dissolve IGEPAL CA-620 in water. Add isopropanol.

Physical Properties:
    pH (as is)   6.4
    pH (1%)   4.7
    Viscosity   10 cps
    Specific Gravity   .98

SOURCE: GAF Corp.: Formulary: Prototype Formulation GAF 5404

WINDOW CLEANER
CONCENTRATE

RAW MATERIALS   % By Weight

IGEPAL CA-620   5.0
Isopropanol   95.0
                                                                     100.0

Manufacturing Procedure
    Mix components together until homogeneous.
    Use dilution 1:10 with water.

Physical Properties
    pH (as is)   8.4
    pH (1%)   5.0
    Viscosity   10 cps
    Specific Gravity   .95

SOURCE: GAF Corp.: Formulary: Prototype Formulation GAF 5405

WINDOW CLEANER, SPRAY

RAW MATERIALS	% By Weight
TRITON QS-30 Surfactant (90%)	0.5
Ammonium Hydroxide (28%)	2.5
Methanol	47.0
Isopropyl Alcohol	50.0
	100.0

Mixing Instructions:
    Add TRITON QS-30 Surfactant to methanol and isopropyl alcohol add ammonium hydroxide last.

Use Dilution: 1 part in 5 parts water.

SOURCE: Rohm and Haas Co.: Lit. Ref. CS-439

WINDOW CLEANER, SPRAY

RAW MATERIALS	% By Weight
TRITON X-100 Surfactant	0.3
Methanol	50.0
Isopropanol	49.7
	100.0

Use Dilution: 1 part in 3 parts water.

SOURCE: Rohm and Haas Co.: Lit. Ref.: CS-427

13. Miscellaneous Cleaners

ABRASIVE CLEANER--HIGH VISCOSITY

RAW MATERIALS   % By Weight

HOSTAPUR SAS 60   6.0
Alkylphenolethoxilate (4 EO)   2.0
Alfol 1214 (Condea)   0.5
Tetra Sodium Pyrophosphate   3.0
CACO3 (DURCAL 40 or CALIBRITE SL)   45.0
Formaldehyde Solution   0.1
Water, Perfume, Oil   AD 100.0%

Production Procedure:
   Dissolve tetra sodium pyrophosphate in water at elevated
temperature. In a separate container the alkylphenol-ethoxilate
and the ALFOL 1214 are mixed thoroughly properly (without lumps).
Add this mixture and then HOSTAPUR SAS to the phosphate solution
at 40-50C, stirring constantly. After the batch is cooled down
to about 30C add formaldehyde, perfume and CaCO3. To guarantee
the stability of the formulation stir again thoroughly after
24 hrs.

SOURCE: Hoechst/Celanese: Formulation D-7003

ALL-SURFACE STEAM CLEANER(2886-086)

RAW MATERIALS   % By Weight

Sodium metasilicate pentahydrate   0.5
Potassium hydroxide (45%)   3.0
Tetrasodium EDTA (40%)   3.0
TRYCOL 6964 POE (9) Nonylphenol   5.0
EMID 6533 Modified Alkanolamide   1.0
TRYFAC 5553 Phosphate Ester   5.0
Water   82.5
                                                          100.0

Blending Procedure:
   Add the water to the blending tank. While mixing add the
ingredients to the blending tank in the order listed. Mix
until uniform.

SOURCE: Emery Chemicals: Specialty Chemicals Formulary:
        Formulation 2886-086

ALKALINE CLEANER

RAW MATERIALS	% By Weight
Water	65
Sodium Gluconate	3
Sodium Silicate	10
45% Potassium Hydroxide	10
MAZER MACOL 212	5
MAZER MAPHOS 60A	5
MAZER MACOL NP 9.5	2

SOURCE: Mazer Chemicals Inc.: Household/Industrial T-20B:
Formulation 20

ASPHALT RELEASE AGENTS

RAW MATERIALS	% By Weight
1L:	
Isopropanol	4.0
EMID 6533 Modified Alkanolamide	20.0
Sodium tripolyphosphate (STPP)	1.0
Dye, fragrance	q.s.
Water	to 100
2L:	
Isopropanol	4.0
EMID 6533 Modified Alkanolamide	15.0
TRYCOL 6964 POE Nonylphenol	15.0
Sodium tripolyphosphate (STPP)	1.0
Dye, fragrance	q.s.
Water	to 100

Blending Procedure:
    Add the water to the blending tank. While mixing, add the
ingredients to the blending tank in the order listed. Mix
until uniform (pH=9)

SOURCE: Emery Chemicals: Specialty Chemicals Formulary:
Formulation 2878-012-1L, 2L

BARBECUE CLEANER: HIGH ALKALINE

RAW MATERIALS	% By Weight
HOSTAPUR SAS 60	15.0
Alkylphenolethoxilate (6 EO)	8.0
Stearic Acid	1.5
Hydrotrope (HOE S 2817)	3.0
NaOH (100%)	10.0
CaCO3	20.0
Water	42.5

Production Procedure:
   Heat alkylphenolethoxilate, hydrotrope and stearic acid
to achieve a transparent melt. Then add NaOH and warm water
(50C). After the neutralization of the stearic acid (approx.
10 min.) gradually add the CaCO3 and homogenize thoroughly.

Instructions:
   Brush the surface of the barbecue with the paste and wait
for about 30 min. Then clean the barbecue brushing under
running water and rinse thoroughly. Attention! Caustic!

SOURCE: Hoechst/Celanese: Formulation D-9003

BOTTLE CLEANER

RAW MATERIALS	% By Weight
Sodium hydroxide (50%)	98
KLEARFAC AA-270 surfactant	0.5
Sodium gluconate	1.5

Suggested use concentration: 2-4 oz. per gallon of water.

SOURCE: BASF Corp.: Formulation #3625

BOTTLE WASH CONCENTRATE

RAW MATERIALS	% By Weight
MIRAWET ASC	2.5
Igepal CO-710	0.5
Sodium Gluconate	20.0
Versene 100	5.0
Water	72.0

Note: This concentrate is to be metered into the caustic
      solution at a ratio of about one part concentrate to
      one hundred parts of 6% NaOH.

SOURCE: Miranol Inc.: MIRANOL Products for Household/Industrial
        Applications: Suggested Formulation

BILGE CLEANER - SOLVENT EMULSIFIER

RAW MATERIALS	% By Weight
Aromatic 150	80.0
TRITON N-57 Surfactant	6.7
TRITON X-114 Surfactant	13.3
	100.0

SOURCE: Rohm and Haas Co.: Lit. Ref. CS-409, CS-421

BILGE CLEANERS

RAW MATERIALS 1M:	% By Weight
EMID 6533 Modified Alkanolamide	10.0
TRYCOL 6964 POE (9) Nonylphenol	5.0
TRYCOL 5940 POE (6) Tridecyl Alcohol	3.0
EMEREST 2665 PEG-600 Dioleate	7.0
Isopropanol	3.0
Water	72.0
	100.0

RAW MATERIALS 2M:	% By Weight
EMID 6533 Modified Alkanolamide	6.0
TRYCOL 6964 POE (9) Nonylphenol	6.0
TRYCOL 5940 POE (6) Tridecyl Alcohol	3.0
EMEREST 2665 PEG-600 Dioleate	10.0
Isopropanol	3.0
Water	72.0
	100.0

RAW MATERIALS 3M:	% By Weight
EMID 6533 Modified Alkanolamide	10.0
TRYCOL 6964 POE (9) Nonylphenol	10.0
TRYCOL 5940 POE (6) Tridecyl Alcohol	3.0
EMEREST 2665 PEG-600 Dioleate	3.0
Isopropanol	3.0
Water	71.0
	100.0

RAW MATERIALS 4M:	% By Weight
EMID 6533 Modified Alkanolamide	7.5
TRYCOL 6964 POE (9) Nonylphenol	7.5
TRYCOL 5940 POE (6) Tridecyl Alcohol	7.5
EMEREST 2665 PEG-600 Dioleate	3.0
Isopropanol	3.0
Water	71.5
	100.0

SOURCE: Emery Chemicals: Specialty Chemicals Formulary: 1-4M

BOTTLE-WASH COMPOUND

RAW MATERIALS	% By Weight
TRITON QS-44 Surfactant	1.25
Sodium Gluconate	1.00
Sodium Hydroxide	97.75
	100.00

Use Dilution: 1 to 4 oz./gal. water

Sodium glucoheptanate can be used as a substitute for Sodium Gluconate.

Lit. Ref: CS-410

BOTTLE WASH, MACHINE-A

RAW MATERIALS	% By Weight
TRITON BG-10 Surfactant	0.5
Sodium Hydroxide (50%)	95.0
Sodium Gluconate	2.5
Water	2.0
	100.0

BOTTLE WASH, MACHINE-B

RAW MATERIALS	% By Weight
TRITON BG-10 Surfactant	0.35
TRITON DF-16 Surfactant	0.15
Sodium Hydroxide (50%)	95.00
Sodium Gluconate	2.50
Water	2.00
	100.00

Mixing Instructions:
   Mix TRITON BG-10 Surfactant (and TRITON DF-16 Surfactant) with water and add caustic solution with care, then sodium gluconate.

Directions for Use:
   Spray hot on bottles in beverage line, rinse well.
   Formulation B is a low-foam version of Formulation A.

Lit. Ref: CS-400, CS-499

SOURCE: Rohm and Haas Co.: Specialty Chemicals: Detergent
        Formulations for Industrial and Institutional Industry

BOWL CLEANER

RAW MATERIALS   % By Weight

VARION TEG   7.5
VARIQUAT 50MC   1.0
HCL (37%) Hydrochloric acid   27.0
Water   qs 100

Mixing Procedure:
   Add the HCl and the 50MC into the water. The TEG is then
added to thicken to required viscosity. More or less may be
required from batch to batch dependent on variation in quality
of industrial muriatic acid.

SOURCE: Sherex: Industrial Formulation 8:1.7

BUTYL CLEANER(HEAVY DUTY DEGREASER)

RAW MATERIALS   % By Weight

NINOL 1281 or 1285   5.0-10.0
NA3, NTA   2.0
Tetrapotassium Pyrophosphate   4.0
STEPANATE X   3.0
Butyl Cellosolve or
 Butyl Carbitol   5.0
Liquid potassium hydroxide (45%)   10.0
Water   Balance

Mixing Procedure:
   Dissolve builders in water. Add butyl cellosolve, STEPANATE
X, NINOL 1281 or 1285, add potassium hydroxide in that order
while mixing.

Properties:
   Appearance   Clear yellow liquid
   pH as is   13.4
   Active, %   20.7-25.7

Use Instructions:
   Use concentration: 2-4 oz/gal

Performance:
   An excellent cleaner/degreaser formula for industrial and
institutional application.

Comments:
   Must not be used on aluminum

STEPAN CO.: Formulation No. 102

BUTYL CLEANER

RAW MATERIALS   % By Weight

MIRANOL CM-SF CONC.   5.0
Dowanol EB   3.7
Igepal CO-630   2.0
Sodium Metasilicate Pentahydrate   4.0
Tetrapotassium Pyrophosphate   4.0
Water   81.3

"BUTYL" CLEANER

RAW MATERIALS   % By Weight

MIRANOL C2M-SF CONC.   5.0
Dowanol EB   10.0
Sodium Carbonate   2.0
Water   83.0

SOURCE: Miranol Inc.: MIRANOL Products for Household/Industrial
        Applications: Formulations

BUTYL CLEANER AND DEGREASER

RAW MATERIALS	% By Weight	CAS Registry Number
Water	83.0	7732-18-5
Trisodium Phosphate	3.0	7601-54-9
Sodium Tripolyphosphate	3.0	7758-29-4
ESI-TERGE HA-20	5.0	Mixture
Potassium Hydroxide	1.0	1310-58-3
Butyl Cellosolve	5.0	111-76-2
	100.0	

Procedure:
    Add in order listed with adequate agitation, allowing each
material to dissolve. Add the ESI-TERGE HA-20 and Butyl Cellosolve. Agitate until clear.

Specifications:
    % Solid   6.5
    % Active   11.0
    pH   12-13
    Viscosity   Water

SOURCE: Emulsion Systems Inc.: Technical Service Bulletin
        Code HA-20-7

CARBON CLEANER

RAW MATERIALS	% By Weight
1.	
methylene chloride	50
xylene	5
cresylic acid	15
DOWANOL DPM glycol ether	15
Igepal CO-630	3
water	12
II.	
methylene chloride	59.5
DOWANOL-P Mix glycol ether	26.5
potassium oleate (80%)	2.4
water	3.4
paraffin wax	8.2

These are dip part cleaners for carbon on engine parts. Allow to soak 10-20 minutes. Water flush off. The water forms a seal to prevent MeCl2 evaporation. This can also be used as an emulsion and brushed on the carbon deposits.
DOWANOL glycol ethers are used as coupling solvents and for cleaning properties.

SOURCE: Dow Chemical U.S.A.: The Glycol Ethers Handbook:
         Suggested Formulations

COSMOLINE REMOVER
(Heavy Oil or Greasy-Wax Remover)

RAW MATERIALS	% By Weight
xylene	30
perchloroethylene	30
Igepal CO-530 Surfactant	20
DOWANOL DPM glycol ether	20

Reduce 1:1 with kerosene.
This can be used to remove protective coatings from automobiles or firearms by a water flush.

SOURCE: Dow Chemical U.S.A.: The Glycol Ethers Handbook:
         Suggested Formulation

CHAIN LUBRICANT

RAW MATERIALS	% By Weight
MIRANOL JEM CONC.	10.0
Caprylic Acid adjusted to pH 11.6 with NaOH, 50%	10.0
Oleic Acid adjusted to pH 11.6 with NaOH, 50%	10.0
Water	70.0

SOURCE: Miranol Inc.: MIRANOL Products for Household/Industrial
Applications: Suggested Formulation

CIP CLEANER

RAW MATERIALS	% By Weight
Sodium tripolyphosphate	47
PLURAFAC RA-40 surfactant	3
Sodium metasilicate pentahydrate	30
Sodium hydroxide	20

Suggested use concentration: 1/4 to 2 oz. per gallon of water.

Formulation #3600

CIP CLEANER

RAW MATERIALS	% By Weight
Sodium tripolyphosphate	35
Sodium carbonate	22
PLURAFAC RA-40 surfactant	3
Sodium metasilicate pentahydrate	40

Suggested use concentration: 1/4 to 2 oz. per gallon of water.

Formulation #3601

SOURCE: BASF Corp.: Cleaning Formulary

COFFEE & TEA MACHINE CLEANER(CONCENTRATE)

RAW MATERIALS	% By Weight
Formic Acid	60.0
VARION AMV SF	10.0
water	30.0

Mixing Procedure: Mix ingredients into water.

SOURCE: Sherex: Industrial Formulation 48:01.1

CONCRETE CLEANER

RAW MATERIALS   % By Weight

Water   83
Ethylene glycol butyl ether   6.5
KLEARFAC AA-270 surfactant   2
ICONOL TDA-8 surfactant   2.5
Sodium metasilicate pentahydrate   3
Sodium tripolyphosphate   3

Use as is

SOURCE: BASF Corp.: Cleaning Formulary: Formulation #3650

CONCRETE CLEANER

RAW MATERIALS   % By Weight

Sodium tripolyphosphate   30
KEARFAC AA-270 surfactant   4
Pine oil   6
sodium metasilicate   60

Suggested use concentration: 2-4 oz. per gallon of water

SOURCE: BASF Corp.: Cleaning Formulary: Formulation #3651

CUTTING AND GRINDING FLUID(LOW FOAM)(2878-012)

RAW MATERIALS   % By Weight

TRYFAC 5555 Phosphate Ester   5.00
Triethanolamine (TEA)   10.00
TRYFAC 5569 Phosphate Ester   6.00
TRYFAC 5576 Phosphate Ester   1.25
Sodium carbonate   0.20
Water   77.55
                                                          100.00

pH = 8.5

Blending Procedure:
    Add the water to the blending tank and warm to 100-120F.
While mixing add the TRYFAC 5555. Continue mixing until it is
completely dissolved and add the remaining ingredients to the
batch tank in the order listed. Mix until uniform.

SOURCE: Emery Chemicals: Specialty Chemicals Formulary:
        Formulation 2878-012

DAIRY MILKSTONE REMOVER

RAW MATERIALS   % By Weight

TRITON X-100 Surfactant	5.0
Phosphoric Acid (85%)	22.0
Water	73.0
	100.0

Mixing Instructions:
  Slowly add phosphoric acid to water with agitation. Add
surfactant and agitate until uniform.
  Weaker acids, such as gluconic or glycolic acids, may be
substituted for phosphoric acid

Directions for Use:
  Soak utensils in a solution of 1/3 to 1 oz. of formula to
1 gallon of water. Brush thoroughly and rinse.

Lit. Ref: CS-427

DAIRY PIPELINE CLEANER(LOW-FOAM)

RAW MATERIALS   % By Weight

TRITON CF-54 Surfactant	5.0
Sodium Hydroxide	10.0
Sodium Silicate (Anhydrous)	30.0
Soda Ash	30.0
Sodium Tripolyphosphate (STPP)	25.0
	100.0

Note:
  TRITON CF-10, TRITON CF-76 or TRITON DF-12 Surfactants may be
substituted for TRITON CF-54 Surfactant.

Use Dilution: 1 to 2 ounces/gallon of water.

Lit. Ref.: CS-60, CS-413, CS-415, CS-436

SOURCE: Rohm and Haas Co.: Specialty Chemicals: Detergent
          Formulations for Industrial and Institutional Industry

DAIRY LINE CLEANER

RAW MATERIALS	% By Weight
VARIQUAT 50MC	10.0
VARION CDG	8.0
Tripotassium Phosphate (TPP)	4.0
Tetrapotassium Pyrophosphate (TKPP)	4.0
Isopropanol	3.0
EDTA	1.5
Water	69.5

Mixing Procedure:
    Dissolve TKPP and TPP into the water. Add the rest of the
ingredients in order. Use moderate stirring to avoid foam.

SOURCE: Sherex: Industrial Formulation 47:05.3

DAIRY LINE CLEANER

RAW MATERIALS	% By Weight
VARION CDG	8.0
VARIQUAT 50MC	10.0
Tri. Potassium Phosphate	4.0
Pot. Pyrophosphate	4.0
Isopropanol	3.0
EDTA	1.5
Water	qs100

Mixing Procedure:
    Dissolve the EDTA and the phosphates in water followed by
the 50MC and CDG. Use the IPA to thin the product.

SOURCE: Sherex: Industrial Formulation 9:05.4.2

DAIRY PIPELINE CLEANER

RAW MATERIALS	% By Weight
Phosphoric acid (75%)	50.0
MAKON NF-5	5.0
Water	45.0

Properties:
    Appearance   Clear liquid
    pH as is   1.5
    Actives, %   42.5
Use Instructions:
    Dilute 2-3 oz/gal and feed into pipeline and recirculate.
Performance:
    Removes milkstone and other soils effectively.
Comments: MAKON NF-5 provides wetting and emulsifying properties
    in addition to low foam.
SOURCE: Stepan Co.: Formulation No. 83

DAIRY CLEANERS*

DAIRY FARM ACID LIQUID**

RAW MATERIALS	% By Weight
NEODOL 25-12	10.0
Phosphoric acid (85%)	57.3
Water	32.7

Properties:
pH	1.3
Phase coalescence temp., F	>185

Use Concentration:
1/4-2 oz/gal.

Blending Procedure:
Add acid to water, surfactant last.

ALKALINE POWDER***

RAW MATERIALS	% By Weight
NEODOL 25-12	5.0
C12 LAS (60%)	3.0
Sodium tripolyphosphate	35.0
Sodium metasilicate, pentahydrate	35.0
Sodium sulfate, decahydrate	22.0

Blending Procedure:
  Mix solid builders and filler thoroughly. Add surfactants slowly while mixing; mix thoroughly.

   * NEODOL surfactants may be used as components of cleaners for food processing equipment. Since the cleaning compound is not considered a food additive, it is not subject to FDA regulations as long as it is followed by a potable water rinse.
  ** Good as milkstone remover and equipment cleaner (manual application)
 *** Use intended for manual application, not for circulation cleaning.

SOURCE: Shell Chemical Co.: The NEODOL Formulary: Formulations

DEGREASER
DEGREASER CONCENTRATES

PREMIUM QUALITY FOR HARD WATER

RAW MATERIALS	% By Weight
NEODOL 91-6	5.0
NEODOL 91-2.5	5.0
Sodium metasilicate, pentahydrate	12.0
EDTA	10.8
TRITON H-66*	12.5
Water, dye, perfume	to 100%

Properties:
    Phase coalescence temp., F   >140

PREMIUM QUALITY FOR MEDIUM-HARDNESS WATER

RAW MATERIALS	% By Weight
NEODOL 91-6	5.0
NEODOL 91-2.5	5.0
Sodium metasilicate, pentahydrate	6.7
EDTA	6.0
Sodium xylene sulfonate (40%)	7.0
Water, dye, perfume	to 100%

Properties:
    Phase coalescence temp., F   >140

HIGH QUALITY CONCENTRATE

RAW MATERIALS	% By Weight
NEODOL 91-6	5.0
NEODOL 91-2.5	5.0
Sodium metasilicate, pentahydrate	10.0
EDTA	4.0
Triton H-66*	10.0
Water, dye, perfume	to 100%

Properties:
    Phase coalescence temp., F   140

Recommended Dilutions:
    Heavy-duty use: 1 part concentrate to 32 parts water (4oz/gal)
    Regular-duty use: 1 part concentrate to 64 parts water
                      (2 oz/gal)
    For high pressure spray system: 1 part concentrate to 10
       parts water (12.8 oz/gal)
    * May replace Triton H-66 with 17.6% sodium xylene sulfonate
       (40%).

SOURCE: Shell Chemical Co.: NEODOL Formulary: Formulations

DEGREASERS
DEGREASER CONCENTRATE

Good Quality Concentrate for High Pressure Spray System

RAW MATERIALS	% By Weight
NEODOL 91-6	3.0
NEODOL 91-2.5	3.0
Sodium metasilicate, pentahydrate	3.0
EDTA	3.0
Sodium xylene sulfonate (40%)	9.0
Water, dye, perfume	to 100%

Properties:
    Phase coalescence temp., F: 140

Recommended Dilutions:
    Heavy-duty use: 1 part concentrate to 32 parts water(4oz/gal)
    Regular-duty use: 1 part concentrate to 64 parts water
        (2oz/gal).
    For high pressure spray system: 1 part concentrate to 10
        parts water (12.8 oz/gal).

SOLVENT DEGREASERS, FLUSH-OFF TYPE

High Quality for Heavy Oils

RAW MATERIALS	% By Weight
NEODOL 91-8	10.0
NEODOL 91-2.5	5.0
NEODOL 25-3	5.0
SHELL SOL 71 or 72*	79.0
Water	1.0

Properties:
    Viscosity, 73F, cps   6
    Phase coalescence temp., F   >176

    *Isoparaffinic solvent, bp 356-401F, Shell Chemical Co.;
    SHELL SOL 140 or Shell Mineral Spirits 145, 150 or 150EC
    can be substituted.

SOURCE: Shell Chemical Co.: NEODOL Formulary: Formulations

DEGREASERS
SOLVENT DEGREASERS, FLUSH-OFF TYPE

High Quality for Regular Oils

RAW MATERIALS	% By Weight
NEODOL 91-6	15.0
NEODOL 91-2.5	5.0
SHELL SOL 71 or 72*	60.0
Butyl OXITOL	18.0
Water	2.0

Properties:
   Viscosity, 73F, cps   8
   Phase coalescence temp., F   >176

   * Isoparaffinic solvent, bp 356-401F, Shell Chemical Co.;
     SHELL SOL 140 or Shell Mineral Spirits 145, 150 or 150EC
     can be substituted.

POWDER DEGREASERS

Caustic, Non-phosphate

RAW MATERIALS	% By Weight
NEODOL 91-6	2.5
NEODOL 91-2.5	2.5
Sodium metasilicate, anhydrous	32.0
Sodium hydroxide, flakes	32.0
Sodium carbonate**	31.0

Non-Caustic, Phosphate

RAW MATERIALS	% By Weight
NEODOL 91-6	2.5
NEODOL 91-2.5	2.5
Sodium metasilicate, anhydrous	30.0
Trisodium phosphate, anhydrous basis	30.0
Sodium carbonate**	35.0

   **May include ~5%w ethylenediamine tetraacetic acid, tetra-
     sodium salt or replace sodium carbonate with trisodium
     phosphate builder for enhanced quality product.

SOURCE: Shell Chemical Co.: NEODOL Formulary: Formulations

DEGREASER

RAW MATERIALS	% By Weight
MIRAWET B	3.5
Dowanol EB	8.5
Sodium Metasilicate Pentahydrate	2.7
Trisodium Phosphate	1.4
Sodium Tripolyphosphate	1.4
Potassium Hydroxide, 45%	1.0
Water	81.5

SOURCE: Miranol Inc.: MIRANOL Products for Household/Industrial
         Applications: Formulation

SPRAY DEGREASER

RAW MATERIALS	% By Weight
Miranol C2M surfactant	3.3
sodium tripolyphosphate	1.4
sodium metasilicate	2.7
trisodium phosphate	1.4
tall oil fatty acids	1.7
potassium hydroxide (45%)	1.0
DOWANOL PM glycol ether	8.5
water	80.0

Excellent degreaser for equipment. DOWANOL PM is a good
grease and oil solvent.

SOURCE: Dow Chemical Co.: The Glycol Ethers Handbook: Formula

HEAVY DUTY SPRAY DEGREASER

RAW MATERIALS	% By Weight
MIRANOL C2M-SF CONC.	3.3
Sodium Tripolyphosphate	1.4
Sodium Metasilicate Pentahydrate	2.7
Trisodium Phosphate	1.4
Actinol FA-2	1.7
Potassium Hydroxide, 45%	1.0
Dowanol EB	8.5
Water	80.0

SOURCE: Miranol Chemical Co.: MIRANOL Products for Household/
         Industrial Applications: Formulation

HIGH FOAM DEGREASER

RAW MATERIALS	% By Weight
MIRANOL H2M CONC.	12.0
Isopropyl Alcohol	4.0
Trisodium Phosphate	0.7
Dowanol EB	2.0
Methocel E4M Premium, 3% solution	10.0
Water	71.3

SOURCE: Miranol Inc.: MIRANOL Products for Household/Industrial
      Applications: Formulas

NON-BUTYL DEGREASER

RAW MATERIALS	% By Weight	CAS Registry Number
Water	91.30	
Soda Ash	1.71	497-19-8
Sodium Metasilicate	2.56	10213-79-3
Tetra Sodium Pyrophosphate	1.29	7720-88-5
ESI-TERGE RT-61	3.41	
	100.00	

Procedure:
   Add in order listed with adequate agitation, allowing each
material to dissolve or desperse completely.
Specifications:

% Solids	8.97
% Active	8.97
pH	12.0-12.5
Viscosity	Low

SOURCE: Emulsion Systems Inc.: Technical Service Bulletin
     Code RT-61-1

NON PHOSPHATE NON BUTYL DEGREASER

RAW MATERIALS	% By Weight	CAS Registry Number
Water	82.3	7732-18-5
Sodium Metasilicate	5.5	10213-79-3
Caustic Potash	2.2	1310-58-3
Versene 100	4.8	64-02-8
ESI-TERGE RT-61	5.2	Not Estabilished
	100.0	

Specifications:

% Solids	15.0	% Active	15.0
pH	13.3	Viscosity	Water

SOURCE: Emulsion Systems Inc.: Technical Service Bulletin RT-61-3

INDUSTRIAL STRENGTH NON BUTYL DEGREASER

RAW MATERIALS	% By Weight	CAS Registry Number
Water	82.6	7732-18-5
Sodium Metasilicate	4.4	10213-79-3
Caustic Potash	1.1	1310-58-3
Tetra Sodium Pyrophosphate	2.2	7320-34-5
Trisodium Phosphate	2.2	7601-54-9
Versene 100	1.1	64-02-8
ESI-TERGE RT-61	5.4	Not Established
ESI-TERGE N-100	1.0	9816-45-9
	100.0	

Procedure:
   Add salts to water and dissolve. Add other ingredients in order mentioned.

Specifications:
% Solids	15.3
% Active	15.3
pH	13.4
Viscosity	Water

SOURCE: Emulsion Systems Inc.: Technical Service Bulletin
          Code RT-61-4

WATER-BASED SOLVENT DEGREASER

RAW MATERIALS	% By Weight
A.	
Ethylene glycol n-butyl ether	10.0
TRYFAC 5553 Phosphate Ester	10.0
Mineral Spirits	8.0
TRYCOL 6961 POE (4) Nonylphenol	3.0
TRYCOL 5940 POE (6) Tridecyl Alcohol	1.0
B.	
Water	64.0
Tetrasodium EDTA (40%)	4.0
	100.0

Blending Procedure:
   Combine the ingredients in Part A in a blending tank and mix until uniform. In a second blending tank, add the ingredients in Part B and mix until uniform. Add Part B to Part A and continue to mix until uniform.

SOURCE: Emery Chemicals: Specialty Chemicals Formulary:
          Formulation 2886-084

DRAIN CLEANER, LIQUID-A

RAW MATERIALS	% By Weight
Water	86.9
ACRYSOL ICS-1 Thickener (30%)	2.5
TRITON X-100 Surfactant	0.1
Sodium Metasilicate Anhydrous	0.5
Sodium Hydroxide (50%)	10.0
	100.0

DRAIN CLEANER, LIQUID-B

RAW MATERIALS	% By Weight
Water	86.87
ACRYSOL ICS-1 Thickener (30%)	2.50
TRITON QS-44 Surfactant (80%)	0.13
Sodium Metasilicate Anhydrous	0.50
Sodium Hydroxide (50%)	10.00
	100.00

Properties: Brookfield Viscosity, cps.
    @ 0.5 rpm: A: 6100   B: 6150
    @ 12 rpm: B: 675   C: 1650
Mixing Procedures: Add ACRYSOL ICS-1 Thickener to the water,
    then the surfactant and metasilicate with adequate agitation.
    Add caustic slowly with high-shear mixing.
Note: A flocculant precipitate may form upon adding the caustic
    solution. It disappears after a few minutes of agitation.
Use Dilution: As prepared

SOURCE: Rohm and Haas Co.: Lit. Ref: CS-410/CS-427/CS-505

DRAIN CLEANER
Liquid Type

RAW MATERIALS	% By Weight
GAFAC RA-600	5.0
Potassium hydroxide	12.0
M-PYROL	3.0
Water	80.0
	100.0

Manufacturing Procedure:
    1. Dissolve GAFAC RA-600 in total amount of water.
    2. Add potassium hydroxide; agitate until mixture is clear.
    3. Add M-PYROL.
Physical Properties:
    pH (as is):   8.4   pH (1%):   11.5
    Viscosity: 10 cps   Specific Gravity: 1.03

SOURCE: GAF Corp.: Formulary: Prototype Formulation GAF 5952

DRY CLEANING SOLUTIONS

RAW MATERIALS	% By Weight

A.

TRITON GR-7M Surfactant (64%)	3.2
Perchloroethylene	95.8
Water	1.0
	100.0

B.

TRITON GR-7M Surfactant (64%)	3.2
Stoddard Solvent	94.8
Water	2.0
	100.0

Use Dilution: As prepared

SOURCE: Rohm and Haas Co.: Specialty Chemicals: Detergent
        Formulations for Industrial and Institutional Industry:
        Lit. Ref: CS-31

DRY CLEANING COMPOUND

RAW MATERIALS	% By Weight

MAZER MAPHOS 76 NA	2
MAZER MAZAWET DOSS (70%)	1
Water	1
Perchloroethylene*	96

* 0.3% DEA may be added at the sacrifice of Perchloroethylene

SOURCE: Mazer Chemicals, Inc.: Household/Industrial T-20B:
        Formula No. 19

DRY CLEANING FLUID

RAW MATERIALS	Parts by Weight

SURFONIC N-60	1-5
Triethylene Glycol (to prevent soil redeposition)	5
Water (maximum for complete solubility)	0.5
Perchloroethylene or	
Trichloroethylene	90

SOURCE: Texaco Chemical Co.: Suggested Formulation

DRY CLEANING FORMULA

RAW MATERIALS	% By Weight
GAFAC PE-510	10.0
GAFAC RS-610	21.0
IGEPAL CO-630	15.0
Potassium hydroxide (50% active)	4.0
Perchlorooethylene	50.0
	100.0

Manufacturing Procedure:
1. Add surfactants together. Mix thoroughly.
2. Add perchloroethylene and potassium hydroxide.

Physical Properties:
pH (as is)	4.3
pH (1%)	3.7
Viscosity	50 cps
Specific Gravity	1.08

SOURCE: GAF Corp.: Formulary: Prototype Formulation GAF 5302

DRY CLEANING FORMULA

RAW MATERIALS	% By Weight
IGEPAL RC-520	30.0
GAFAC RS-610	30.0
NEKAL WT-27	7.5
Hexylene glycol	10.0
Stoddard solvent	18.5
Potassium hydroxide (50% active)	4.0
	100.0

Manufacturing Procedure:
1. Add GAFAC RS-610 to IGEPAL RC-520. Mix thoroughly.
2. Add NEKAL WT-27, hexylene glycol, stoddard solvent, mixing well after each addition.
3. Add potassium hydroxide slowly to main batch.

Physical Properties:
pH (as is)	4.3
pH (1%)	3.4
Viscosity	150 cps
Specific Gravity	1.00

SOURCE: GAF Corp.: Formulary: Prototype Formulation GAF 5303

EMULSIFIABLE SOLVENT CLEANERS
(GARAGE FLOOR CLEANER)

RAW MATERIALS	% By Weight
Solvent	86
Emulsifier	9
Coupling Solvent	5

Any combination of solvent-emulsifier-coupling solvent can be used.
Solvent:
    kerosene
    perchloroethylene
    DOWCLENE EC
    mineral spirits
Emulsifier:
    Igepal CO-630
    Atlas G-3300
    Igepal CO-630
Coupling Solvent:
    DOWANOL PM
    DOWANOL DPM
The concentrate should be diluted with 3-4 parts of solvent and applied to surface to be cleaned. Allow to soak briefly and rinse off with water.
Heavy lubricating oils and greases can be removed from metal surfaces or floors with a water flush.

SOURCE: Dow Chemical Co.: The Glycol Ethers Handbook: Formulas

FOAMING BUTYL CLEANER

RAW MATERIALS	% By Weight	CAS Registry Number
Water	65.70	
Trisodium Phosphate	4.35	7601-54-9
Sodium Metasilicate	4.35	10213-79-3
Potassium Hydroxide (45%)	3.50	1310-58-3
ESI-TERGE 320	3.60	52276-83-2
ESI-TERGE DDBSA	5.25	27176-87-0
ESI-TERGE SXS	4.50	1300-72-7
Butyl Cellosolve	8.75	111-76-2
	100.00	

Specifications:
    % Solids   19-20
    % Active   27-28
    pH   13-13.5
    Viscosity   Water Like

SOURCE: Emulsion Systems Inc.: Technical Service Bulletin 320-8

FOAM MARKER CONCENTRATES
HIGH VISCOSITY

RAW MATERIALS	% By Weight
NEODOL 25-3S (60%)	17.5
Alcohol sulfate (30%)	31.5
Cocobetaine (30%)	20.0
FADEA	4.0
Isopropyl alcohol	5.0
Water	22.0

Properties:
Viscosity, 73F, cps	418	
Phase coalescence temp., F	>176	
pH	9.7	
Foam Ht, mm	~93	

LOW VISCOSITY

RAW MATERIALS	% By Weight
NEODOL 25-3S (60%)	33.5
Cocobetaine (30%)	15.0
FADEA	4.0
Isopropyl alcohol	5.0
Water	42.5

Properties:
Viscosity, 73F, cps	55	
Phase coalescence temp., F	>176	
pH	9.7	
Foam Ht, mm	~125	

Blending Procedure:
Add NEODOL 25-3S last, slowly with good stirring.

SOURCE: Shell Chemical Co.: NEODOL Formulary: Suggested Formulas

FOOD PROCESSING EQUIPMENT CLEANER, SPRAY

RAW MATERIALS	% By Weight
TRITON CF-76 Surfactant	2.0
Sodium Tripolyphosphate	50.0
Sodium metasilicate, anhydrous	31.0
Soda Ash	15.0
Sodium Dichloro-s-triazinetrione (CDB Clearon)	2.0
	100.0

Use Dilution: 2-4 oz./gallon water

SOURCE: Rohm and Haas Co.: Lit. Ref. CS-413

GEL RUST REMOVER

RAW MATERIALS   % By Weight

MIRANOL JS CONC.   4.0
Sodium Hydroxide, 50%   50.0
Sodium Gluconate   6.0
Versene 100   1.0
Water   39.0

Note: This product is liquid when prepared. It will set to a
      rigid gel in approximately 12 hours.

SOURCE: Miranol Inc.: MIRANOL Products for Household/Industrial
        Applications: Formulations

GRAFFITI REMOVER

RAW MATERIALS   % By Weight

MAZER MAFO 13   10.0
Potassium Hydroxide (45%)   40.0
MAZER MACOL 212   5.0
MAZER MACOL 48   5.0
Sodium Gluconate   5.0
Sodium Xylene Sulfonate (40%)   5.0
Water   30.0

SOURCE: Mazer Chemicals, Inc.: Household/Industrial T-20B:
        Formulation 10

GRAFFITI REMOVER

RAW MATERIALS   % By Weight

1.
DOWANOL PM glycol ether   50
DOWANOL DB glycol ether   30
isopropanol   20

II.
DOWANOL PM glycol ether   40
DOWANOL DPM glycol ether   30
pine oil   10
AEROTHENE TT Solvent   20

   Removes all types of graffiti except some of the wax type
markings. Formulation I will not attack painted surfaces.

SOURCE: Dow Chemical U.S.A.: The Glycol Ethers Handbook:
        Formulations

GLASS/BOTTLE LIQUID CLEANER COMPOUND

RAW MATERIALS	% By Weight
NEODOL 91-6	2.0
Sodium gluconate	2.5
Sodium hydroxide (50%)	20.0
Triton H-66	8.0
Water, dye, perfume	to 100%

Properties:

Viscosity, 73F, cps	7
Phase coalescence temp., F	>176
Clear point, F	32
pH	13.2

SOURCE: Shell Chemical Co.: NEODOL Formulary: Formulation

GRILL CLEANER

RAW MATERIALS	% By Weight
TRITON X-100 Surfactant	5.0
TRITON H-66 Surfactant (50%)	4.0
Sodium Metasilicate (anhydrous)	3.0
Tetrapotassium Pyrophosphate (TKPP)	3.0
Dipropylene Glycol Methyl Ether (Dowanol DPM)	5.0
TAMOL SN Dispersant	1.0
Water	79.0
	100.0

Use Dilution: 1 to 2 ounces/gallon water.

SOURCE: Rohm and Haas Co.: Specialty Chemicals: Detergent
Formulations for Industrial and Institutional Industry:
Lit. Ref: CS-71/CS-427/CS-433

HEAVY DUTY CLEANER*

RAW MATERIALS	% By Weight	CAS Registry Number
Water	84.0	7732-18-5
Trisodium Phosphate	5.0	7601-54-9
Sodium Metasilicate	5.0	10213-79-3
ESI-TERGE HA-20	6.0	Mixture
	100.0	

Procedure:
Add in order listed with adequate agitation. Allow all powder
to dissolve before adding ESI-TERGE HA-20. Agitate until clear.
Specifications:

% Solids	16	% Active	16
pH	12.5-13.5	Viscosity	Medium

* To convert to a wax stripper, 3-5% ammonia or monoethanolamine is added to this cleaner.

SOURCE: Emulsion Systems Inc.: Technical Service Bulletin HA-20-6

HEAVY-DUTY CLEANER--DRY

RAW MATERIALS	Parts By Weight
SURFONIC N-95	15-20
Sodium Carbonate	40
Sodium Sulfate	30
Tetrasodium Pyrophosphate	10
Sodium Metasilicate	5

SOURCE: Texaco Chemical Co.: Suggested Formulation

HEAVY DUTY CLEANER AND DEGREASER(2878-024)

RAW MATERIALS	% By Weight
TRYCOL 5941 POE (9) Tridecyl Alcohol	3.0
TRYFAC 5559 Phosphate Ester	6.0
Triethanolamine (TEA)	3.0
Tetrapotassium pyrophosphate (TKPP)	2.0
Sodium metasilicate pentahydrate	1.5
Ethylene glycol n-butyl ether	10.0
Water	75.5
	100.0

Blending Procedure:
    Add the water to the blending tank and heat the water to
140-150F. While mixing, add the ingredients to the water in
the order listed. Completely dissolve the TKPP before adding
the metasilicate. Cool to room temperature before packaging.

Use Dilution:
    Dilute 2-25 parts of product to 25 parts of water.

SOURCE: Emery Chemicals: Specialty Chemicals Formulary:
        Formulation 2878-024

HEAVY DUTY CONCRETE CLEANER

RAW MATERIALS	% By Weight
MIRANOL C2M-SF CONC.	4.0
Tetrapotassium Pyrophosphate	6.0
Trisodium Phosphate	1.0
Sodium Metasilicate Anhydrous	2.0
Triton X-100	2.0
Carbitol	2.0
Water	83.0

SOURCE: Miranol Inc.: MIRANOL Products for Household/Industrial
        Applications: Suggested Formulation

HEAVY DUTY HOUSEHOLD CLEANER(2878-001)

RAW MATERIALS	% By Weight

4D:

Potassium nydroxide (45%)	2.5
Tetrasodium EDTA (40%)	4.0
Sodium metasilicate pentahydrate	1.5
TRYFAC 5559 Phosphate Ester	3.0
TRYCOL 5941 POE (9) Tridecyl Alcohol	2.0
TRYCOL 6965 POE (11) Nonylphenol	5.0
Dye, fragrance, etc.	as desired
Water	to 100

5D:

Potassium hydroxide (45%)	2.5
Tetrasodium EDTA (40%)	4.0
Sodium metasilicate pentahydrate	1.5
TRYFAC 5559 Phosphate Ester	3.0
TRYCOL 6965 POE (11) Nonylphenol	5.0
Dye, fragrance, etc.	as desired
Water	to 100

Blending Procedure:
   Add the water to the blending tank. While mixing, add the ingredients to the blending tank in the order listed. Stir until uniform.

Use Dilution:
   2-4 ounces (1/4 to 1/2 cup) per gallon of water.

SOURCE: Emery Chemicals: Specialty Chemicals Formulary:
        Formulation 2878-001-4D, 5D

HEAVY DUTY LIQUID STEAM CLEANER

RAW MATERIALS	% By Weight
MIRANOL J2M CONC.	1.0
Potassium Hydroxide, 45%	55.0
Kasil #1	32.0
Gluconic Acid, 50%	4.0
Phosphoric Acid, 75%	8.0

Mix in order listed.

SOURCE: Miranol Inc.: MIRANOL Products for Household/
        Industrial Applications: Formulations

HEAVY-DUTY PAINT STRIPPER

RAW MATERIALS	% By Weight
MIRAWET ASC	3.0
Potassium Hydroxide (45%)	30.0
Sodium Gluconate	4.0
Water	63.0

SOURCE: Miranol Inc.: MIRANOL Products for Household/Industrial
Applications: Formulations

HIGH ALKALINE CLEANER

RAW MATERIALS	% By Weight	CAS Registry Number
Water	77.5	
Sodium Tripolyphosphate	5.0	7758-29-4
Trisodium Phosphate	5.0	7601-54-9
Sodium Metasilicate	5.0	10213-79-3
ESI-TERGE 320	7.5	52276-83-2
	100.0	

Procedure:
    Add in order listed with adequate agitation, allowing powders
to dissolve completely before adding ESI-TERGE 320.

Specifications:
% Solids	22.5
% Active	22.5
pH	12.5-13.5
Viscosity	Low

SOURCE: Emulsion Systems Inc.: Technical Service Bulletin 320-1

HIGH ALKALINE CLEANER

RAW MATERIALS	% By Weight	CAS Registry Number
Water	76.0	7732-18-5
*Potassium Hydroxide (90%)	6.0	1310-58-3
Trisodium Phosphate	6.0	7601-54-9
Sodium Metasilicate	6.0	10213-79-3
ESI-TERGE 330	6.0	Not Established
	100.0	

Specifications:
% Solids	22.5	% Activity	22.5
pH	12.5-13.5	Viscosity	Water

* Formula may be adjusted for use of 45% Potassium Hydroxide
SOURCE: Emulsion Systems Inc.: Technical Service Bulletin 330-1

HIGH PRESSURE CLEANER
Liquid, Clear

RAW MATERIALS	% By Weight
Nonylphenolethoxylat (+6 EO)	2,5
Nonylphenolethoxylat (+8 EO)	2,5
HOE S 2817	3,0
KOH (85%)	5,0
Thermphos NW	10,0
Sodium-meta-silicate	10,0
Water	67,0

Production procedure:
   Dissolve HOE S 2817 and KOH in warm water (60C). Than add Nonylphenolethoxylate, Thermphos NW and Sodium-meta-silicate.

Tests:
   pH-value   13,1
   Viscosity   13 mPas
   Stability (-5C)   clear
   Freeze and Thaw Test i.o.

SOURCE: Hoechst/Celanese: Formulation E-1004

HIGH PRESSURE CLEANER
Acid

RAW MATERIALS	% By Weight
HOSTAPUR SAS 30	12.0
Nonylphenolethoxilate (8 EO)	1.0
Phosphoric Acid, 85%	7.0
Citric Acid Monohydrate	3.0
Water	AD100%

SOURCE: Hoechst/Celanese: Suggested Formulation

HIGH PRESSURE STEAM CLEANER

RAW MATERIALS	% By Weight
VARION EP AMVSF	10.0
Sodium Hydroxide (33%)	25.0
Monoethanolamine (MEA)	5.0
Trilon B (EDTA Sodium Salt Solution)	9.0
Water	qs100

Mixing Procedure:
   Dissolve MEA, Glycol and Trilon into water then the AMVSF and follow with the 33% Sodium Hydroxide.

SOURCE: Sherex: Industrial Formulation 2:05.6

HOT PLATE/GRILL CLEANER

RAW MATERIALS	% By Weight
Sodium metasilicate, anhydrous	22.0
CHEELOX NTA-Na3	1.0
GAFAC RA-600	7.0
GAFAMIDE CDD-518	2.0
Sodium xylene sulfonate (40% active)	10.0
Sodium hydroxide	1.0
Water	57.0

Manufacturing Procedure:
1. Dissolve sodium metasilicate, anhydrous.in water.
2. Add CHEELOX NTA-NA3 and sodium xylene sulfonate. Mix thoroughly.
3. Add GAFAC RA-600, GAFAMIDE CDD-518 and sodium hydroxide, mixing after each addition.
4. Filter product.

Physical Properties:
pH (as is)	13.0
pH (1%)	11.6
Viscosity	40 cps
Specific Gravity	1.08

SOURCE: GAF Corp.: Formulary: Prototype Formulation GAF 5654

INDUSTRIAL CLEANER
Alkaline

RAW MATERIALS	% By Weight
HOSTAPUR SAS 60	6.0
Laurylethersulfate, 28%	3.0
Sodium Tripolyphosphate	4.0
Sodium Metasilicate	8.0
Butylglycol	5.0
Water	74.0

SOURCE: Hoechst/Celanese: Suggested Formulation

LIQUID HIGH PRESSURE CLEANER CONCENTRATE

RAW MATERIALS	% By Weight
MIRANOL C2M-SF CONC.	20.0
Tetrapotassium Pyrophosphate	22.0
Sodium Metasilicate Pentahydrate	10.0
Sodium Benzoate	1.0
Sodium Xylene Sulfonate, 40%	1.0
Dowanol EM	2.0
Water	44.0

Procedure:
    Dissolve TKPP in water at 80C, then cool to 60C. Separately weigh out the sodium metasilicate pentahydrate, then pour the TKPP solution at 60C over the metasilicate. Stir until dissolved, then add in order: sodium benzoate, sodium xylene sulfonate, MIRANOL C2M-SF CONC., and Dowanol EM.
an alternate formulation:

MIRANOL C2M-SF CONC.	20.00
Starso	70.00
Sodium Benzoate	0.25
Water	9.75

SOURCE: Miranol Inc.: MIRANOL Products for Household/Industrial
        Applications: Suggested Formulations

LIQUID HIGH PRESSURE CONCENTRATE

RAW MATERIALS	% By Weight
VARION AMSKSF40	20.0
Potassium Pyrophosphate	20.0
Sodium Metasilicate	10.0
Sodium Benzoate	1.0
Sodium Xylene Sulfonate (nxs 40)	1.0
Dowanol EM	2.0
Water	qs 100

SOURCE: Sherex: Industrial Formulation 44:05.6

LIQUID ACID DAIRY SANITIZER & CLEANER

SOURCE	% By Weight
MIRANOL C2M-SF CONC. (adjusted to pH 7.0)	13.27
Triton X-100	6.63
Quaternary Ammonium Germicide, 50%	13.00
Glycolic Acid	17.00
Water	50.10

Note: When diluted to contain from 100 to 200 ppm quaternary
        in water this formula has excellent hard water tolerance.

SOURCE: Miranol Inc.: MIRANOL Products: Sugggested Formulation

LIQUID RUST REMOVER

RAW MATERIALS	% By Weight
MIRANOL JEM CONC.	2.0
Potassium Hydroxide, 45%	75.0
Triethanolamine	12.0
Water	11.0

LIQUID CAUSTIC CLEANER

RAW MATERIALS	% By Weight
MIRANOL J2M CONC. or	
MIRANOL J2M-SF CONC.	2.0
Potassium Hydroxide, 45%	10.0
Kasil #1	50.0
Water	38.0

LOW FOAM SPRAY WASHING COMPOUND

RAW MATERIALS	% By Weight
MIRANOL J2M-SF CONC.	5.3
Potassium Hydroxide, 45%	44.5
Tetrapotassium Pyrophosphate, 60%	18.0
Water	32.2

SOURCE: Miranol Inc.: MIRANOL Products for Household/Industrial
Applications: Suggested Formulations

LIPSTICK STAIN REMOVER

RAW MATERIALS	% By Weight
Igepal CO-630 Surfactant	0.1
DOWANOL PM glycol ether	11.0
DOWANOL DB glycol ether	23.0
naphtha (60-100C)	65.9

Also excellent for removing magic marker ink, grease ballpoint
pen marks and pencil marks from counter tops, desks, school lock-
ers and other hard surfaces.

LEATHER, VINYL, PLASTIC CLEANER

RAW MATERIALS	% By Weight
Igepal CO-630	10.0
DOWANOL PM glycol ether	5.0
isopropanol	2.5
amyl acetate	1.0
water	81.5

SOURCE: Dow Chemical U.S.A.: The Glycol Ethers Handbook: Formulas

LEATHER CLEANER

RAW MATERIALS	% By Weight
VARION 2C	15.0
Siponate A246L (Alpha Olefin Sulfonate)	15.0
PVP - Iodine	10.0
VARIQUAT 50MC	2.0
VARAMIDE ML-1	3.0
water	55.0

Mixing Procedure:
    Disperse molten ML-1 into the 2C; add the PVP - Iodine and
50MC into water followed by the Siponate. Then finally the
2C/ML-1 blend. Adjust pH to 6.5 with 50% citric acic.

SOURCE: Sherex: Industrial Formulation 50:01.1

MEDIUM DUTY STEAM CLEANER(2886-088)

RAW MATERIALS	% By Weight
Sodium metasilicate pentahydrate	2.0
Potassium hydroxide (45%)	3.0
Tetrasodium EDTA (40%)	2.5
Triethanolamine (TEA)	3.0
TRYCOL 5941 POE (9) Tridecyl Alcohol	2.5
TRYFAC 5553 Phosphate Ester	6.0
Water	81.0

Blending Procedure:
    Add the water to the blending tank. While mixing, add the
ingredients to the blending tank in the order listed. Mix until
uniform.
Use Dilution:
    Dilute 1 part of the formulated product with 50-100 parts of
water before using.

SOURCE: Emery Chemicals: Specialty Chemicals Formulary:
        Formulation 2886-088

OIL FIELD APPARATUS CLEANER

    Tne following formula is recommended for rig-washing in oil-
fields:

RAW MATERIALS	Parts By Weight
SURFONIC N-95	20
Alcohol	5
Water	80

    Add more alcohol to reduce foaming or gelling. Allow the
formula to remain one-half hour or more and then rinse.

SOURCE: Texaco Chemical Co.: Suggested Formulation

MEDIUM DUTY NON-CAUSTIC STEAM CLEANER(2886-112)

RAW MATERIALS   % By Weight

TRYCOL 5941 POE (9) Tridecyl Alcohol   3.0
TRYCOL 6965 POE (11) Nonylphenol   2.0
TRYFAC 5553 Phosphate Ester   5.0
Tetrasodium EDTA (40% aq)   5.0
Sodium carbonate   5.0
Dye, fragrance   as desired
Water   to 100

Blending Procedure:
    Add the water to the blending tank. While mixing, add the
ingredients to the blending tank in the order listed. Mix
until uniform.

Use Dilution:
    Dilute 1 part of the formulated product with 50-150 parts
water.

SOURCE: Emery Chemicals: Specialty Chemicals Formulary:
        Formulation 2886-112

PAINT BRUSH CLEANER

RAW MATERIALS   % By Weight

I.
metnylene chloride   73
water   2
Renex 36 surfactant   5
DOWANOL PM glycol ether   20

II.
xylene   90
DOWANOL P-Mix glycol ether   6
Miramine OC   4

SOURCE: Dow Chemical U.S.A.: The Glycol Ethers Handbook: Formulas

PRODUCE PEELING FORMULA

RAW MATERIALS   % By Weight

MIRANOL C2M-SF CONC.   1.0
Carbitol   1.0
Potassium Hydroxide, 45%   43.0
Water   55.0

SOURCE: Miranol Inc.: MIRANOL Products for Household/Industrial
        Applications: Formulation

PORCELAIN CLEANER

RAW MATERIALS	% By Weight	CAS Registry Number
Water	76.00	
Trisodium Phosphate	2.00	7601-54-9
ESI-TERGE B-15	5.25	61789-19-3
Tall Oil Fatty Acid	1.75	8002-26-4
Kaopolite SF	15.00	7631-86-9
	100.00	

Procedure:
1. Dissolve Trisodium Phosphate in water.
2. When solution clears up add ESI-TERGE B-15 and mix well until clear.
3. Add tall oil fatty acid and mix until emulsion clears.
4. Add Kaopolite SF and mix until homogeneous blend is formed.

Specifications:
Activity	24%
pH	9.5-10.5
Viscosity	680 cps
Specific Gravity	1.068

SOURCE: Emulsion Systems Inc.: Technical Service Bulletin B-15-3

HEAVY DUTY POT AND PAN CLEANER(2878-118)

RAW MATERIALS	% By Weight
Tetrasodium EDTA (40%)	2.5
TRYCOL 5967 POE (12) Lauryl Alcohol	15.0
EMID 6533 Modified Alkanolamide	15.0
Citric Acid (50%) (to pH 7.0-8.0)	q.s.
Dye, fragrance, preservative	as desired
Water	to 100

Blending Procedure
   Add the water to the blending tank. While mixing, add the remaining ingredients in order listed. Mix until uniform.

SOURCE: Emery Chemicals: Specialty Chemicals Formulary: Formulation 2878-118

POWDERED CAUSTIC CLEANER

RAW MATERIALS	% By Weight
MIRANOL J2M CONC. or	
MIRANOL J2M-SF CONC.	1.0- 2.0
Sodium Gluconate	6.0- 6.0
Sodium Hydroxide	93.0-92.0

SOURCE: Miranol Inc.: MIRANOL Products for Household/Industrial
Applications: Formulation

RUST REMOVER-LIQUID-ACID

RAW MATERIALS	% By Weight
NEODOL 91-6	0.5
Phospnoric acid (85%)	12.0
Butyl DIOXITOL	2.0
Water	to 100%

Properties:
Viscosity, 73F, cps	5
Phase coalescence temp., F	172
pH	1.8

RUST REMOVER-LIQUID-ALKALINE

RAW MATERIALS	% By Weight
NEODOL 91-6	4.0
Sodium gluconate	1.0
Potassium hydroxide (45%)	35.0
Triethanolamine	5.0
Triton H-66	7.0
Water	to 100%

Properties:
Viscosity, 73F, cps	10
Phase coalescence temp., F	>176
pH	14

Use Instructions:
The cleaners may be diluted with water or used "as is" on
ferrous and non-ferrous metal surfaces.

SOURCE: Shell Chemical Co.: NEODOL Formulary: Suggested Formulas

SOLVENT EMULSION, SLUDGE AND CARBON CLEANER

RAW MATERIALS   % By Weight

Kerosene   57.5
Cresylic Acid   20.0
Heavy Aromatic Naphtha   10.0
o-Dichlorobenzene   10.0
TRITON X-102 Surfactant   2.5
                                                           100.0

SOURCE: Rohm and Haas Co.: Lit. Ref: CS-407

SOLVENT CLEANER

RAW MATERIALS   % By Weight

SURFONIC N-95   5
SURFONIC N-40 (and/or N-10)   5
Kerosine (deodorized mineral spirits or Stoddard
 Solvent is best)   90

Mix in any manner.

Applications:
    Apply by spray, wiping, or brushing, whichever is best for
the application.
    1. Clothes - tar and grease removal before washing.
                     Will also remove most paints.
    2. Paint brush cleaner
    3. Driveway cleaner
    4. Engine cleaner
    5. Hand cleaner
    6. Tar and grease remover for car finishes
    7. Remove nydrophobic soils from almost any surface
    Apply to water-free surfaces, allow to stand for 1-5 minutes
for action, and rinse with water.

SOURCE: Texaco Chemical Co.: Suggested Formulation

SPOTTING LIQUID FOR DRY CLEANING

RAW MATERIALS   % By Weight

TRITON X-114 Surfactant   50.0
Perchloroethylene   40.0
Isopropyl Alcohol   5.0
Water   5.0
                                                           100.0

Directions for Use: Before dry cleaning, spray or moisten cotton,
                    cotton/polyester, permanent press fabrics.

SOURCE: Rohm and Haas Co.: Lit. Ref: CS-409

SPOT CLEANER

These materials take advantage of the solvency of DOWANOL glycol ethers towards water, soap and chlorinated solvents. The finished fluid contains, in a one-phase formulation, the solvent, such as perchloroethylene, which will dissolve grease stains, plus a soap and water for emulsifying and removing water-soluble stains.

All Purpose Emulsion

RAW MATERIALS	% By Weight
I.	
water	6
sodium hydroxide	4
DOWANOL DPM glycol ether	12
triethanolamine	10
oleic acid	47
perchloroethylene	21

Mix in order listed.
  Cleans oily and greasy spots from fabrics and carpets. Not good for lipstick stains.

II.	
Igepal CO-630	1.5
DOWANOL DPM glycol ether	15.5
isopropanol	13.0
perchloroethylene	30.0
VM & P naphtha	40.0

SOURCE: Dow Chemical U.S.A.: The Glycol Ethers Handbook: Formulas

SYNTHETIC CLEANER

RAW MATERIALS	% By Weight	CAS Registry Number
Water	89.0	
Trisodium Phosphate	0.8	7601-54-9
Sodium Tripolyphosphate	1.6	7758-29-4
Potassium Hydroxide (90%)	0.3	1310-58-3
Tall Oil Fatty Acid	3.3	8002-26-4
ESI-TERGE B-15	5.0	
	100.0	

Procedure:
  Add in order listed with adequate agitation. Allow all powders to dissolve before adding tall oil fatty acid and ESI-TERGE B-15. Agitate until clear.

Specifications:

% Solids	11.0
% Active	11.0
pH	8.5-9.5
Viscosity	Heavy

SOURCE: Emulsion Systems Inc.: Technical Service Bulletin B-15-1

STEAM CLEANER, LIQUID

RAW MATERIALS   % By Weight

TRITON H-55 Surfactant (50%)	2.0
TRITON X-114 Surfactant	0.4
Potassium Hydroxide	14.1
Sodium silicate (1.8 SiO2/Na2O 38%)	40.0
Tetrapotassium Pyrophosphate (TKPP)	17.2
Water	26.3
	100.0

Use Dilution: 1 part in 20 parts hot water.

Lit. Ref: CS-409, CS-433

STEAM CLEANER POWDER
(Phosphate-Free)

RAW MATERIALS   % By Weight

TRITON N-101 Surfactant	2.0
Sodium Metasilicate (Anhydrous)	43.3
Tetrasodium Ethylenediaminetetraacetate (Versene 100)	17.0
Sodium Citrate Dihydrate	37.7
	100.0

Use Dilution: 1 oz. per gal. hot water.

Lit. Ref: CS-408

STEAM CLEANING COMPOUND

RAW MATERIALS   % By Weight

Sodium Hydroxide	30.0
Borax	30.0
Soda Ash	20.0
Disodium Phosphate	15.0
TRITON QS-15 Surfactant	5.0
	100.0

Directions for Use: Add 2 oz./gallon water heated to 140F.

Lit. Ref: CS-417

SOURCE: Rohm and Haas Co.: Specialty Chemicals: Detergent
        Formulations for Industrial and Institutional Industry

STEAM CLEANER

RAW MATERIALS	% By Weight	CAS Registry Number
Water	93.00	7732-18-5
Sodium Hydroxide or	5.00	1310-73-2
Potassium Hydroxide 97%		
ESI-TERGE 330	1.00	Not Established
Versene 100	1.00	64-02-8
	100.00	

Procedure:
   Add in order listed.

Specifications:
Solids	6.5-7.0
Active	6.5-7.0
pH	13.0-13.5
Viscosity	Water Like

Formulation Note:
   Solids may be increased proportionally or sodium metasilicate may be added. For 5 parts combination of alkali or
silicates use 1 part of ESI-TERGE.

SOURCE: Emulsion Systems Inc.: Technical Service Bulletin 330-4

STEAM CLEANER

RAW MATERIALS	% By Weight	CAS Registry Number
Water	93.0	7732-18-5
Potassium Hydroxide (90%)	5.0	1310-58-3
ESI-TERGE HA-20	1.0	Mixture
Versene	1.0	64-02-8
	100.0	

Procedure:
   Add salts to water and dissolve.
   Add other ingredients in order mentioned.

Specifications:
% Solids	5.9
% Active	5.9
pH	13.25-13.75
Viscosity	Low

SOURCE: Emulsion Systems Inc.: Technical Service Bulletin
        HA-20-4

STEAM CLEANERS
Steam/Spray

RAW MATERIALS	% By Weight
NEODOL 91-6	0.4
Sodium Silicate (37.5%)	15.2
Tetrapotassium pyrophosphate	17.2
Potassium Hydroxide (45%)	31.1
Triton H-66	2.0
Water	to 100%

Properties:
Viscosity, 73F, cps	11
Phase coalescence temp., F	>176
pH	>14

Powder

RAW MATERIALS	% By Weight
NEODOL 91-6	2.5
DDBSA (98%)	4.2
Sodium metasilicate, anhydrous	15.0
NTA	3.0
Sodium gluconate	4.0
Sodium hydroxide, flakes	10.0
Sodium carbonate	63.0

Recommended Dilution:
    6 oz/gal.

Non-Phosphate Concentrate

RAW MATERIALS	% By Weight
NEODOL 91-6	2.0
C12 LAS (60%)	5.0
Potassium hydroxide (45%)	25.0
NTA	1.0
Sodium gluconate	4.0
Sodium xylene sulfonate (40%)	12.0
Water	to 100%

Properties:
Viscosity, 73F, cps	7
Phase coalescence temp., F	>176
pH	>14

Recommended Dilution:
    1 part cleaner to 50 parts water (2.6oz/gal).

SOURCE: Shell Chemical Co.: NEODOL Formulary: Suggested Formulas

STEAM CLEANING COMPOUNDS

RAW MATERIALS	% By Weight
I.	
MIRANOL C2M-SF CONC.	15.0-15.0
Sodium Metasilicate Pentahydrate	50.0-20.0
Water	35.0-65.0
II.	
MIRANOL C2M-SF CONC.	15.0-15.0
Sodium Metasilicate Pentahydrate	30.0-20.0
Potassium Hydroxide, 45%	11.0-11.0
Water	44.0-54.0
III.	
MIRANOL C2M-SF CONC.	15.0
Sodium Metasilicate Pentahydrate	20.0
Potassium Hydroxide, 45%	22.0
Water	43.0

Procedure:
    Dissolve the sodium metasilicate in the water at 70C. Cool to 40C and add the MIRANOL C2M-SF CONC. The liquid potassium hydroxide may then be added at any time before use.

RAW MATERIALS	% By Weight
MIRANOL C2M-SF CONC.	15.0
Starso	62.0
Potassium Hydroxide, 45%	10.0
Water	13.0

SOURCE: Miranol Inc.: MIRANOL Products for Household/Industrial
        Applications: Suggested Formulations

TEXTILE SOFTENER

RAW MATERIALS	% By Weight
MIRANOL DM	71.0
Paraffin Wax	10.0
Tween 60	1.0
Span 80	1.0
Nonylphenoxypoly (ethyleneoxy) Ethanol	1.0
Water	16.0

Note: This formulation is applied to fabrics such as shirting
      to enhance needle penetration and sewing speed.

SOURCE: Miranol Inc.: MIRANOL Products for Household/Industrial
        Applications: Suggested Formulation

TANK CLEANERS

Soak Tank

RAW MATERIALS	% By Weight
NEODOL 25-3S (60%)	5.0
Potassium silicate (29.1%)	15.0
Potassium hydroxide (45%)	10.0
EDTA	1.9
Sodium xylene sulfonate (40%)	5.0
Water	to 100%

Properties:
    Viscosity, 73F, cps   6
    Phase coalescence temp., F   >176
    pH   >14

Fuel Oil (Bunker C) Tank

RAW MATERIALS	% By Weight
NEODOL 91-6*	4.3
NEODOL 25-3	1.4
NEODOL 45-7*	2.9
Tetrapotassium pyrophosphate	1.4
Sodium metasilicate, pentahydrate	2.0
Sodium xylene sulfonate (40%)	6.0
Water	to 100%

Properties:
    Viscosity, 73F, cps   7
    Phase coalescence temp., F   131
    pH   >12

* NEODOL 91-8 can be used in place of NEODOL 91-6, and NEODOL 25-7 or NEODOL 23-6.5 can be used in place of NEODOL 45-7, with only very minor changes in physical properties.

SOURCE: Shell Chemical Co.: NEODOL Formulary: Suggested Formulas

Section II
Automotive Cleaners

14. Car and Truck Washes

AUTO SHAMPOO

RAW MATERIALS   % By Weight

MIRANOL C2M-SF CONC.   5.0
MIRATAINE CBC   10.0
Dodecylbenzene Sulfonic Acid   12.0
Sodium Hydroxide (50%)   3.0
Igepal CO-630   3.0
Cedemide CX   3.0
Water   64.0

SOURCE: Miranol Inc.: MIRANOL Products for Household/Industrial
        Applications: Formulation

AUTO SHAMPOO (NON-STREAKING)

RAW MATERIALS   % By Weight

VARION AMKSF 40   15.0
SLES (30%)   10.0
Water   qs 100

Mixing Procedure:
   Add the AMKSF to water first followed by the SLES.

SOURCE: Sherex: Industrial Formulation 3:05.2.1

AUTOMOBILE FOAMING SPRAY

RAW MATERIALS   % By Weight

CARSPRAY 700   25% @110F
Mineral Seal Oil   25%
Tap Water   50%

   The product imparts sheen and a temporary protection to the
finish.

SOURCE: Sherex: CARSPRAY 700 Carnuaba Foamer

BOAT WASH & CAR CLEANER

RAW MATERIALS   % By Weight

Water   39
MAZER MAZAMIDE 80   11
MAZER MACOL NP 9.5   5
MAZER MAZON 60T   45

SOURCE: Mazer Chemicals, Inc.: Automotive Formularies T-20A: 1

CAR WASH DETERGENT POWDER

RAW MATERIALS	% By Weight
Sodium Tripolyphosphate (STPP)	85.0
TRITON X-114 Surfactant	15.0
	100.0

Directions for Use:
  Premix 0.5 lb. free-flowing powder with 1 gal. water.
  For car wash, use 1 gal. of solution per 25 gal. water.

Lit. Ref.: CS-409

CAR WASH DETERGENT POWDERS

RAW MATERIALS	% By Weight
A.	
TRITON X-114 Surfactant	7.5
TRITON X-45 Surfactant	7.5
Sodium Metasilicate (Anhydrous)	20.0
Carboxymethylcellulose (4000 cps.)	1.0
Soda Ash (Light Density)	10.0
Sodium Hexametaphosphate	5.0
Sodium Tripolyphosphate (STPP)	49.0
	100.0
B.	
TRITON X-114 Surfactant	7.5
TRITON X-45 Surfactant	7.5
Sodium Metasilicate (Anhydrous)	5.0
Carboxymethylcellulose (4000 cps.)	1.0
Soda Ash (Light Density)	10.0
Sodium Hexametaphosphate	10.0
Sodium Tripolyphosphate (STPP)	59.0
	100.0

Use Dilution: 2 oz./car

Directions for Use:
   Charge and dissolve into concentrate tank and meter into wash-ing stream. Excellent soil removal in high pressure automatic carwash installation. Add 3% TAMOL SN Dispersant to improve detergency and dispersion of light soils. Add 3% lanolin wax to give wax polish.

Lit. Ref.: CS-409/CS-403

SOURCE: Rohm and Haas Co.: Specialty Chemicals: Detergent
        Formulations for Industrial and Institutional Industry

CAR WASH, LIQUID

RAW MATERIALS	% By Weight
TRITON N-101 Surfactant	12.0
Sodium Linear Alkylate Sulfonate (60%)	16.0
Lauricdiethanolamide	2.0
Water	70.0

Use Dilution: 2 oz./gal. water

Lit. Ref: CS-408

CAR WASH LIQUID

RAW MATERIALS	% By Weight
A.	
TRITON X-100 Surfactant	20
TRITON X-301 Surfactant (20%)	10
Water	70
	100
B.	
TRITON X-100 Surfactant	15
Sodium Linear Alkylate Sulfonate (60%)	15
Water	70
	100
C.	
TRITON X-100 Surfactant	12
Sodium Linear Alkylate Sulfonate (60%)	16
Lauricdiethanolamide	2
Water	70
	100
D.	
TRITON X-102 Surfactant	12
Sodium Linear Alkylate Sulfonate (60%)	23
Lauricdiethanolamide	3
Water	62
	100

Directions for Use:
1 oz./6 to 8 gal. water, or more concentrated if heavily soiled. Use 1 oz./7.5 to 10 gal. water for D. For manual cleaning. Rinse after cleaning.
   A is excellent detergent for difficult soils.
   B is least expensive.
   C has best foam stability.
   D is more concentrated, has good detergency, foam stability

Lit. Ref: CS-33/CS-407/CS-427

SOURCE: Rohm and Haas Co.: Specialty Chemicals: Detergent
        Formulations for Industrial and Institutional Industry

CAR WASH, THICKENED LIQUID

RAW MATERIALS	% By Weight
TRITON X-114 Surfactant	15.0
TRITON X-301 Surfactant	5.0
ACRYSOL ICS-1 Thickener (30%)	4.0
Water	75.9
Sodium Hydroxide (50%)	0.1
	100.0

Viscosity: LVT Brookfield, 3 rpm - Spindle #2 - 4400 cps.
            6 rpm - Spindle #2 - 3600 cps.

Use Dilution: 1 to 2 oz./gal. water

Lit. Ref.: CS-33/CS-409/CS-504/CS-505

SOURCE: Rohm and Haas Co.: Specialty Chemicals: Detergent
        Formulations for Industrial and Institutional Industry

LIQUID CAR WASH

RAW MATERIALS	% By Weight
Water	34
MAZER MAZON 60T	30
MAZER MACOL OP-10 SP	10
MAZER MAZAMIDE 80	6
MAZER MACOL 212	10
Methanol	10

LIQUID CAR WASH

RAW MATERIALS	% By Weight
MAZER MAZON 41	40.0
MAZER MACOL NP 9.5	5.0
Water	55.0

LIQUID CAR WASH CONCENTRATE

RAW MATERIALS	% By Weight
Avanel S-30	15.0
MAZER MACOL NP 9.5	5.0
MAZER MAZON 41	40.0
Water	40.0

SOURCE: Mazer Chemicals, Inc.: Automotive Formularies T-20A

CAR WASH
Powdered Type

RAW MATERIALS   % By Weight

IGEPAL CO-710	10.0
Sodium tripolyphosphate	50.0
Sodium metasilicate 5H2O	5.0
Sodium carbonate (lt. density)	35.0
	100.0

Manufacturing Procedure:
1. Mix IGEPAL CO-710 with sodium tripolyphosphate.
2. Add sodium metasilicate 5-H20 and sodium carbonate.

Physical Properties:
   pH (1%)   11.2
   Specific Gravity   .73

SOURCE: GAF Corp.: Formulary: Prototype Formulation GAF 5452

CAR WASH
Liquid Type

RAW MATERIALS   % By Weight

Alkylbenzene sulfonic acid	6.0
Sodium xylene sulfonate	11.5
IGEPAL CO-710	3.0
Tetrapotassium pyrophosphate (60% active)	25.0
Potassium hydroxide (50% active)	2.5
Water	52.0
	100.0

Manufacturing Procedure:
1. Dissolve alkylbenzene sulfonic acid in water.
2. Add sodium xylene sulfonate, IGEPAL CO-710, tetrapotassium pyrophosphate and potassium hydroxide, mixing well after each addition.
3. Filter product.

Physical Properties:
   pH (as is)   9.5
   pH (1%)   9.9
   Viscosity   10 cps
   Specific Gravity   1.05

SOURCE: GAF Corp.: Formulary: Prototype Formulation GAF 5453

CAR WASH LIQUID CONCENTRATES

High Quality

RAW MATERIALS	% By Weight
NEODOL 25-3S (60%)	15.0
NEODOL 91-6	8.0
C12 LAS (60%)	30.0
FADEA	5.0
Ethanol	3.0
Water, dye, perfume	to 100%

Properties:

Viscosity, 73F, cps	213
Phase coalescence temp., F	>176
Clear point, F	28
pH	10.3

Good Quality

RAW MATERIALS	% By Weight
NEODOL 25-3S (60%)	13.9
NEODOL 91-6	7.0
C12 LAS (60%)	27.0
FADEA	3.0
Ethanol	3.0
Water, dye, perfume	to 100%

Properties:

Viscosity, 73F, cps	235
Phase coalescence temp., F	>176
Clear point, F	28
pH	9.2

Economy

RAW MATERIALS	% By Weight
NEODOL 25-3S (60%)	8.3
NEODOL 91-6	5.0
C12 LAS (60%)	16.7
FADEA	3.0
Ethanol	2.0
Water, dye, perfume	to 100%

Properties:

Viscosity, 73F, cps	240
Phase coalescence temp., F	>176
Clear point, F	36
pH	9.4

SOURCE: Shell Chemical Co.: NEODOL Formulary: Formulations

CAR WASH LIQUID CONCENTRATE

Generic

RAW MATERIALS	% By Weight
NEODOL 25-3S (60%)	6.7
NEODOL 91-6	4.0
C12 LAS (60%)	13.3
FADEA	2.0
Water, dye, perfume	to 100%

Properties:
Viscosity, 73F, cps	68
Phase coalescence temp., F	>176
Clear point, F	32
pH	9.5

CAR WASH POWDERS

Premium Quality

RAW MATERIALS	% By Weight
NEODOL 91-6	10.0
Sodium tripolyphosphate	80.0
Sodium metasilicate, anhydrous	10.0

Blending Procedure:
   Mix solid builders thoroughly.
   Add NEODOL 91-6 slowly while mixing, mix thoroughly.

Good Quality

RAW MATERIALS	% By Weight
NEODOL 91-6	5.0
DDBSA (98%)	5.0
Sodium carbonate	42.0
Sodium tripolyphosphate	40.0
Sodium metasilicate, anhydrous	8.0

Blending Procedure:
   Adsorb DDBSA onto sodium carbonate, then mix with other
solid builders thoroughly. Add NEODOL 91-6 slowly while mixing,
mix thoroughly.

SOURCE: Shell Chemical Co.: NEODOL Formulary: Formulations

CAR WASH CONCENTRATE WITH ALCOHOL ETHOXYLATE

RAW MATERIALS	% By Weight
Water, D.I.	36.0
DESONOL SE	15.0
DESONATE 60-S	30.0
PETRO LBA Liquid	6.0
DESONIC 1036	8.0
Varamide MA-1	5.0

Blending Procedure:
    Blend ingredients in the order listed.

Typical Properties:
    Viscosity = 315 cps
    % Actives = 43.0

CAR WASH CONCENTRATE WITH ALCOHOL ETHOXYLATE

RAW MATERIALS	% By Weight
Water, D.I.	43.3
DESONOL SE	13.9
DESONATE 60-S	27.8
PETRO LBA Liquid	5.0
DESONIC 1036	7.0
Varamide MA-1	3.0

Blending Procedure: Blend ingredients in the order listed.

Typical Properties:
    Viscosity = 335 cps
    % Actives = 37.5

MEDIUM-COST CAR WASH

RAW MATERIALS	% By Weight
DESONOL SE	10.0
Cocamide DEA	4.0
PETRO LBA Powder	2.0
Formaldehyde, Inhibited	0.1
Sodium Chloride	q.s.
Water (D.I.), Perfume, Dye	83.9
Adjust pH = 7.5-8.0	

Blending Procedure: Blend the ingredients in the order listed.
    Sodium Chloride is used to adjust the viscosity of the
    finished product.

SOURCE: DeSoto, Inc.: Suggested Formulations

CAR WASH DETERGENT(2878-115)

RAW MATERIALS	% By Weight
Water	to 100
Caustic soda (50% sodium hydroxide)	3.3
Dodecylbenzene sulfonic acid (DDBSA)	13.0
TRYCOL 5943 POE (12) Tridecyl Alcohol	2.0
EMID 6500 Cocoamide MEA	2.0
EMERSAL 6453 Sodium Laureth Sulfate	15.0
Sodium xylene sulfonate (40%) (SXS)	3.0
Citric acid (50%) (to pH 6.5-7.5)	q.s.
Dye and fragrance	as desired

Blending Procedure:
   To the batch tank, add the water, caustic soda and DDBSA.
At this point, the pH should be greater than 7.0. If not,
immediately add more caustic soda. Heat the batch to 150-170F.
Add the TRYCOL 5943 and EMID 6500. Mix until the EMID 6500
has completely dissolved. Cool to 110F and add the remaining
ingredients.

Note:
   The viscosity of the finished product can be adjusted by
increasing or decreasing the SXS.

SOURCE: Emery Chemicals: Specialty Chemicals Formulary:
         Formulation 2878-115

HAND CAR WASH

RAW MATERIALS	% By Weight	CAS Registry Number
Water	84.4	7732-18-5
Sodium Tripolyphosphate	2.2	7758-29-4
ESI-TERGE T-60	9.8	27323-41-7
ESI-TERGE HA-20	3.6	Not Established
	100.0	

Procedure:
   Add salts to water and dissolve. Add other ingredients in
order mentioned.

Specifications:
% Solids	11.6
% Active	11.6
pH	8.8
Viscosity	41 cps

SOURCE: Emulsion Systems Inc.: Technical Service Bulletin T-60-3

CAR SHAMPOO
Transparent, Medium Viscosity

RAW MATERIALS	% By Weight
HOSTAPUR SAS 60	24.0
Laurylethersulfate-Na (28%)	22.0
Cocofattyacid Diethanolamide	3.0
NaCl	2.0
Water, Preservative	49.0

Production Procedure:
    Dissolve HOSTAPUR SAS, Laurylethersulfate, cocofattyacid diethanolamide and preservative in water, stirring constantly. Then adjust the viscosity with NaCl.

Formulation F-1001

CAR SHAMPOO
Acid

RAW MATERIALS	% By Weight
HOSTAPUR SAS 60	3.0
Nonylphenolethoxilate (6 EO)	2.0
Phosphoricacidesters/Phosphoric Acid-Mixture	20.0
Water	AD100%

CAR CLEANER
High Foaming

RAW MATERIALS	% By Weight
HOSTAPUR SAS 60	20.0
Isotridecylalcohol Ethoxilate (8 EO)	5.0
Formaldehyde Solution	0.1
Water	AD 100%

Formulation F-1002

SOURCE: Hoechst/Celanese: Suggested Formulations

CAR SPRAY RINSE

RAW MATERIALS	% By Weight
CARNAUBA SPRAY 200	25
Deodorized Kerosene or	
Carspray Oil	25
Water	50

    The formulation effectively causes water to bead on the clean car surface so that it may be quickly blown dry. In the final carnauba-based formulation the product imparts sheen and a temporary protection to the finish.

CAR SPRAY RINSE

RAW MATERIALS	% By Weight
CARSPRAY #2 CONCENTRATE	25
Deodorized Kerosene or	
Carspray Oil	25
Water	50

    The product imparts temporary protection to the car finish.

CAR SPRAY RINSE

RAW MATERIALS	% By Weight
CARSPRAY 300	20
Butyl Cellosolve	5
Deodorized Kerosene or	
Carspray Oil	25
Water	50

AUTOMATIC CAR WASH

    Formulator should dilute the product 4 parts to 1 part CARSPRAY CW. Effective use level is 0.8-1.5 oz. per auto, with standard applications equipment. Product gives excellent cosmetic effects with foaming.

MANUAL CAR WASH - HOME USE

    Formulator should dilute the product 50/50 with water. Recommended home use will be 2.0-3.0 oz. of formula to 2-3 gallons of water. Rinse auto with clear water, then wash and rinse. Towel dry metal surface and glass.

SOURCE: Sherex: Products for Car Spray Formulation: Formulas

CAR SPRAY WASH

RAW MATERIALS   % By Weight

Sodium Carbonate (Soda Ash), Light Density   15.0
Sodium Tripolyphosphate, Light Density   40.0
DESOPHOS 5AP   2.0
Sodium Tripolyphosphate, Hexahydrate   10.0
DESONIC 9N   2.0
Trisodium Phosphate, Crystal   7.8
Sodium Metasilicate, Pentahydrate   7.0
DESODET 1239   6.0
Sodium Tripolyphosphate, Powder   10.0
*Blue Dye   0.2

Blending Procedure:
    Premix Soda Ash and Tripoly, Light Density together; Slowly
add DESOPHOS 5AP while mixing. Mix until product is uniform
in appearance; Add Tripoly, Hexahydrate; Add DESONIC 9N while
mixing; mix until product looks dry; Add Trisodium Phosphate
and Metasilicate; Add DESODET 1239 while mixing; Add Tripoly,
Powder.
* Comment: If Blue Dye is used, premix with DESONIC 9N.

SOURCE: DeSoto, Inc.: Formulation

WAND-TYPE CAR WASH(2886-082)

RAW MATERIALS   % By Weight

TRYCOL 5941 POE (9) Tridecyl Alcohol   3.0
TRYCOL 5951 POE (6) Decyl Alcohol   1.0
TRYFAC 5553 Phosphate Ester   5.0
Tetrasodium EDTA (40%)   4.0
Dye and fragrance   as desired
Water   to 100

Blending Procedure:
   Add the water to the blending tank. While mixing, add the
ingredients to the blending tank in the order listed. Stir
until uniform.

SOURCE: Emery Chemicals: Specialty Chemicals Formulary:
        Formula 2886-082

MEDIUM DUTY TRUCK AND RIG WASH, POWDER

RAW MATERIALS   % By Weight

Sodium Tripolyphosphate, Light Density	42.0
Sodium Carbonate (Soda Ash)	9.0
DESOPHOS 5AP	3.0
DESONIC 9N	4.2
Trisodium Phosphate, Anhydrous	6.0
Sodium Metasilicate, Anhydrous	14.0
UDET 950	2.0
DESODET 1239	3.8
Sodium Tripolyphosphate, Powder	16.0

Blending Procedure:
   Premix Tripoly and Soda Ash together in mixer; add DESOPHOS 5AP slowly while mixing. Mix well; Add DESONIC 9N and mix until uniform; Add Trisodium Phosphate and Metasilicate; Add UDET 950; Slowly add DESODET 1239 and mix until uniform; Add Tripoly, Powder.

SOURCE: DeSoto, Inc.: Formulation

HEAVY DUTY TRUCK CLEANER (2887-053)

RAW MATERIALS   % By Weight

Trisodium Phosphate	8.0
Ethylene glycol n-butyl ether	8.0
TRYCOL 6952 POE (15) Nonylphenol	2.0
TRYCOL 6961 POE (4) Nonylphenol	1.0
Water	81.0
	100.0

Blending Procedure:
   Charge the water to the blending tank. While mixing, add the ingredients to the blending tank in the order listed. Stir until uniform.

Use Dilution:
   Dilute to desired strength.

SOURCE: Emery Chemicals: Specialty Chemicals Formulary:
        Formulation 2887-053

<u>TRUCK WASH</u>
Liquid Type

RAW MATERIALS   % By Weight

IGEPAL CO-630   3.0
ALIPAL CO-436   3.5
GAFAMIDE CDD-518   2.0
CHEELOX NTA-NA3   3.5
Ethylene glycol monobutyl ether   3.0
Water   85.0
                                                             100.0

Manufacturing Procedure:
   1. Dissolve surfactants IGEPAL CO-630, ALIPAL CO-436, and
      GAFAMIDE CDD-518 in water.
   2. Add CHEELOX NTA-NA3. Mix well.
   3. Add ethylene glycol monobutyl ether.

Physical Properties:
   pH (as is)   9.9
   pH (1%)   9.6
   Viscosity   10 cps
   Specific Gravity   1.01

SOURCE: GAF Corp.: Formulary: Prototype Formulation GAF 5455

<u>TRUCK WASH</u>
High Pressure Spray Type

RAW MATERIALS   % By Weight

IGEPAL CO-660   3.0
CHEELOX NTA-NA3   11.0
Sodium xylene sulfonate   1.5
Water   84.5
                                                             100.0

   * EMULPHOGENE DA-630 may be substituted for IGEPAL CO-660.

Manufacturing Procedure:
   1. Dissolve sodium xylene sulfonate in water.
   2. Add surfactant. Mix thoroughly. Add CHEELOX NTA-NA3.

Physical Properties:
                                   CO-660   DA-630
   pH (as is)   10.3   11.4
   pH (1%)   10.1   10.3
   Viscosity   10 cps   10 cps
   Specific Gravity   1.02   1.02

SOURCE: GAF Corp.: Formulary: Prototype Formulation GAF 5456

TRUCK WASH (LAS/HV9/AMINE OXIDE)

RAW MATERIALS	% By Weight
REWORYL NKS 50 (alkyl benezene sulfonate)	7.0
REWOPOL HV9 (nonoxyl-9)	3.5
REWORYL NXS 40 (Na. xylene sulfonate 40%)	2.75
Versene 100	2.0
Sodium Metasilicate	0.75
Isopropanol	4.0
Sodium Hydroxide	1.8
VAROX 185E	1.0
Water	77.2

Mixing Procedure:
   Dissolve the Versene and metasilicate in all the water, and
then pourin the two Surfactants with minimal stirring to avoid
excessive foaming. Add the Isopropanol to thin the solution,
followed by the sodium hydroxide. When the solution is clear,
add the VAROX 185E to thicken (and improve the foam quality).

SOURCE: Sherex: Industrial Formulation 37:05.2.1

LIQUID TRUCK WASHES

RAW MATERIALS	% By Weight
MIRANOL C2M-SF CONC.	6.0
Dowanol EB	2.0
Igepal CO-630	3.0
Potassium Hydroxide, 45%	4.0
Tall Oil	3.4
Tetrapotassium Pyrophosphate	4.0
Sodium Metasilicate Pentahydrate	4.0
Water	73.6

RAW MATERIALS	% By Weight
MIRANOL C2M-SF CONC.	4.0
CEDEPHOS FA 600M	2.0
Nitrilotriacetic Acid, Trisodium Salt (40% Solution)	17.5
Tetrapotassium Pyrophosphate	1.0
Igepal CO-630	2.0
Potassium Hydroxide, 45%	6.0
Sodium Metasilicate Pentahydrate	2.0
Dowanol EB	2.5
Water	63.0

SOURCE: Miranol Inc.: MIRANOL Products for Household/
         Industrial Applications: Formulations

TRUCK - CAR WASH
Liquid Concentrate

RAW MATERIALS	Parts By Weight
SURFONIC N-85 or N-95	68
Propylene glycol monobutyl ether	28
Potassium Hydroxide	0.5
Water	3.5

SOURCE: Texaco Chemical Co.: Formulation

LIQUID SPRAY CLEANER FOR ALUMINUM TRUCKS

RAW MATERIALS	% By Weight
Water	50.0
Phosphoric Acid (75%)	20.0
Citric Acid	10.0
MAZER MACOL 48	4.0
Ammonium Bifluoride	3.0
Avanel S-30	3.0

Procedure: Dilute with five (5) Parts water.

SOURCE: Mazer Chemicals, Inc.: Automotive Formularies T-20A: 13

VEHICLE WASH

RAW MATERIALS	% By Weight
Water	87
PLURAFAC D-25 surfactant	8
Tetrapotassium pyrophosphate	2
Ammonium hydroxide (28%)	3

Suggested use concentration: 1/4 to 3 oz. per gallon of water
Formulation #3700

VEHICLE WASH

RAW MATERIALS	% By Weight
Water	74
Propylene glycol	6
Sodium xylene sulfonate (40%)	6
PLURAFAC D-25 surfactant	4
Sodium metasilicate pentahydrate	3
Sodium gluconate	4
Trilon B powder (EDTA, tetrasodium salt)	3

Suggested use concentration: 1/4 to 3 oz. per gallon of water
Formulation #3701

SOURCE: BASF Corp.: Cleaning Formulary

15. Whitewall Tire Cleaners

AUTOMOBILE WHITEWALL TIRE CLEANER

RAW MATERIALS	% By Weight
NEODOL 91-6*	2.0
Sodium metasilicate, pentahydrate	8.7
Trisodium phosphate, dodecahydrate	8.7
C12 LAS (60%)	1.6
Sodium xylene sulfonate (40%)	7.0
Water, dye and perfume	to 100%

Properties:
Viscosity, 73F, cps	6
Phase coalescence temp., F	148
pH	13.6

* NEODOL 91-8 can be used in place of NEODOL 91-6 with only very minor changes in physical properties.

SOURCE: Shell Chemical Co.: NEODOL Formulary: Formulation

FOAMY WHITEWALL TIRE CLEANER, LIQUID

RAW MATERIALS	% By Weight
Water, D.I.	68.0
Sodium Tripolyphosphate	2.0
Trisodium Phosphate, Crystal	1.0
Liquid Caustic Potash 45%	12.0
Sodium Silicate 40 Be'	6.0
DESONIC 9N	1.0
Alkali Surfactant	1.0
DESODET 1239	4.0
DESONOL SE	2.0
Butyl Cellosolve	2.0
PETRO BA Liquid	1.0

Blending Procedure: Blend ingredients in the order listed.

Typical Properties:
Specific Gravity	1.087
Wt/Gal	9.06
pH, as is	13.19
Dilution Ratio	1:20 at Wand

SOURCE: DeSoto, Inc.: Formulation

"WHITE LIGHTNING" TIRE CLEANER

RAW MATERIALS	% By Weight
Water	84
MAZER MAZON 60T	3
MAZER MAPHOS 66H	3
MAZER MACOL OP-10 SP	2
Sodium Metasilicate Pentanydrate	2
Sodium Hydroxide Beads	3
MAZER MACOL 212	3

WHITE WALL TIRE BLEACH

RAW MATERIALS	% By Weight
MAZER MAZON 60T	3
MAZER MACOL OP-10 SP	2
MAZER MAZON 40	1
Sodium Tripolyphosphate	2
Sodium Hydroxide (50%)	3
MAZER MACOL 212	10
Water	79

SOURCE: Mazer Chemicals, Inc.: Automotive Formularies T-20A,
        Formulations 2, 3

WHITE SIDEWALL TIRE CLEANER(2886-084)

RAW MATERIALS	% By Weight
Sodium metasilicate pentahydrate	2.0
Potassium hydroxide (45%)	3.0
Tetrasodium EDTA (40%)	10.0
Triethanolamine (TEA)	2.0
TRYCOL 5940 POE (6) Tridecyl Alcohol	1.0
Ethylene glycol n-butyl ether	5.0
TRYFAC 5556 Phosphate Ester	5.0
Dye and fragrance	as desired
water	to 100

Blending Procedure:
   Add the water to the blending tank. While mixing, add the
ingredients to the blending tank in the order listed. Stir
until uniform.
Use Dilution:
   Dilute 1 part of formula to 4-10 parts water before using.

SOURCE: Emery Chemicals: Specialty Chemicals Formulary:
        Formulation 2886-084

WHITE-WALL TIRE CLEANER

RAW MATERIALS	% By Weight
Sodium metasilicate, anhydrous	5.0
Trisodium phosphate, anhydrous	5.0
TRITON QS-44 Surfactant (80%)	0.9
TRITON X-100 Surfactant	1.3
Water	87.8
	100.0

Use Dilution: Use as prepared.

SOURCE: Rohm and Haas Co.: Lit. Ref: CS-410/CS-427

WHITE WALL TIRE CLEANER

RAW MATERIALS	Parts By Weight
Ethylene Glycol Monobutyl Ether	5
SURFONIC N-95	15
Isopropanol	35
Sodium Sesquicarbonate	2
Water	43

Dissolve the sodium sesquicarbonate in the water; then add in order the isopropanol, ethylene glycol monobutyl ether, and SURFONIC N-95. This formulation may be used as is or be made into a paste by the addition of an inorganic filler and white pigment

SOURCE: Texaco Chemical Co.: Formulation

WHITE WALL TIRE CLEANER
Liquid Type

RAW MATERIALS	% By Weight
IGEPAL CA-630	2.0
GAFAC RA-600	6.0
M-PYROL	2.0
Sodium metasilicate, anhydrous	3.0
Sodium tripolyphosphate	5.0
Potassium hydroxide	1.5
Water	80.5
	100.0

Physical Properties:
pH (as is)	13.3
pH (1%)	10.7
Viscosity	10 cps
Specific Gravity	1.03

SOURCE: GAF Corp.: Formulary: Prototype Formulation GAF 5451

16. Miscellaneous Cleaners

ACIDIC CLEANER FOR ALUMINUM TRUCKS

RAW MATERIALS	% By Weight
MIRANOL CS CONC.	10.0
Phosphoric Acid, 75%	50.0
Dowanol EB	10.0
Ammonium Bifluoride	1.0-3.0
Water	QS

Use of MIRANOL JS CONC. will give a low-foaming cleaner.

SOURCE: Miranol Inc.: MIRANOL Products for Household/Industrial
Applications: Formulation

CARBURETOR CLEANER

RAW MATERIALS	Parts by Volume
AEROTHENE MM	20.60 gallons
DOWANOL P-MIX or DOWANOL EB	19.20 gallons
Water	10.20 gallons
Potassium oleate (83% active)	25.80 gallons
Sodium nitrite	00.85 gallons

SOURCE: Dow Chemical U.S.A.: The Glycol Ethers Handbook:
Formulations

AUTOMOBILE WINDSHIELD CLEANERS

RAW MATERIALS	% By Weight
AEROSOL MA-80	3.5
Surfonic N-95	0.5
Isopropanol	51.0
Water	45.0

The above is diluted 5 oz. to 1 gallon of water for use. In wintertime the recommended dilution is 10 oz. per gallon. Freezing is prevented by the isopropanol. This formulation leaves a small amount of residue on the unwiped portion of the windshield on drying. There are several ways of preventing these visible residues.
1. Use a mixture of volatile solvents with or without surface agents. A suggested formulation is:

RAW MATERIALS	% By Weight
2-Ethyl Hexanol	5
AEROSOL OT-75%	1
Isopropanol	47
Water	47

The above is diluted 5-10 oz. per gallon of water, and will not freeze at a higher level of concentration.

2. Using surface active agents, but including small amounts of non-volatile solvents which will give a transparent film on the glass with the surface active agent. Non-volatile solvents having approximately the same refractive index of glass should be selected. A suggested formula is:

RAW MATERIALS		Parts by Weight
AEROSOL MA-80%	(80% basis)	2.5
AEROSOL OT-75%	(75% basis)	3.3
Diethylene Glycol		4.4
Isopropanol		40.0
n-Butyl-p-Hydroxy Benzoate		0.01
Methyl-p-Hydroxy Benzoate		0.01
Water		49.78

The above concentrate is used at 5 oz. per Gallon (10 Oz./ Gal. in Winter). The bactericides are present to keep the window cleaning solution free of slime and tubidity due to bacteria.

RAW MATERIALS	% By Weight
AEROSOL OT-75%	3.5
Dibutylphtnalate or Glycerine	0.5-2.0
Isopropanol	48
Water	46-48

The above is diluted 5-10 oz. per gallon of water and will not freeze at the higher level of concentration.

SOURCE: Angus Chemical Co.: Suggested Formulations

ENGINE DEGREASER CONCENTRATE

RAW MATERIALS	% By Weight
MAZER MAZON 71	25.0
MAZER MAZAWET 77	1.0
Kerosene	74.0

Procedure:
    Dilute 1 part concentrate to 4-9 parts Kerosene at time of use.

SOURCE: Mazer Chemicals, Inc.: Automotive Formularies T-20A: 7

EXTERIOR RAILCAR CLEANER

RAW MATERIALS	% By Weight
MIRANOL CM-SF CONC.	10.0
Dowanol EB	3.8
Tetrapotassium Pyrophosphate	4.0
Sodium Metasilicate Pentahydrate	6.0
Sodium Hydroxide Flake	3.0
Water	73.2

SOURCE: Miranol Inc.: MIRANOL Products for Household/Industrial
        Applications: Formulation

POWDERED HUBCAP CLEANER

RAW MATERIALS	% By Weight
MAZER MACOL 25	3.5
MAZER MAZAWET 77	0.5
MAZER MAZON 41	8.0
Tetrasodium Pyrophosphate	20.0
Sodium Tripolyphosphate	40.0
Ammonium Bifluoride	8.0
Sodium Bicarbonate	10.0
Sodium Metasilicate (Pentahydrate)	10.0

Procedure:
    Use 2-5 ounces to one (1) gallon of water.

SOURCE: Mazer Chemicals, Inc.: Automotive Formularies T-20A: 9

SOLVENT EMULSION CLEANER

RAW MATERIALS	% By Weight
Water	88.13
ACRYSOL ICS-1 Polymer (30%)	1.67
Deodorized Kerosene	10.00
Sodium Hydroxide (50%)	0.20
	100.00

Brookfield Viscosity, cps.
```
   @ 0.5 rpm - 21,000
   @  12 rpm -  2,300
   pH          9.2
```

Add ingredients in listed order. High-shear mixing is necessary to disperse kerosene.

SOURCE: Rohm and Haas Co.: Specialty Chemicals: Detergent Formulations for Industrial and Institutional Industry: Lit. Ref: CS-504

VINYL TOP CLEANER
For Automobiles

RAW MATERIALS	% By Weight
VEEGUM	1.0
Carboxymethylcellulose	0.3
M-PYROL	5.0
IGEPAL CO-660	8.0
Tetrapotassium pyrophosphate (60% active)	3.3
Water	82.4
	100.0

Manufacturing Procedure:
1. Heat water to 85C. Add VEEGUM and carboxymethylcellulose. Stir for one hour maintaining temperature at 85C.
2. Cool to room temperature and add IGEPAL CO-660, M-PYROL and tetrapotassium pyrophosphate. Mix thoroughly after each addition.

Physical Properties:
pH (as is)	10.0
pH (1%)	6.5
Viscosity	100 cps
Specific Gravity	1.01

SOURCE: GAF Corp.: Formulary: Prototype Formulation GAF 5476

WINDSHIELD WASHER CLEANERS

High Quality

RAW MATERIALS	% By Weight
NEODOL 23-6.5	1.5
NEODOL 25-3S (60%)	2.0
Isopropyl alcohol	47.5
Water, dye	to 100%

Properties:

Viscosity, 73F, cps	8
pH	7.7

Recommended Dilution:
    Dilute 1 part cleaner with 2 parts water.

Good Quality

RAW MATERIALS	% By Weight
NEODOL 23-6.5	1.0
Isopropyl alcohol	39.0
Water, dye	to 100%

Properties:

Viscosity, 73F, cps	7
pH	6.5

Recommended Dilution:
    Dilute 1 part cleaner with 1 part water.

Winter Use

RAW MATERIALS	% By Weight
NEODOL 91-8	1.0
Butyl OXITOL	5.0
Propylene glycol	14.0
Isopropyl alcohol	80.0

Properties:

Viscosity, 73F, cps	7
pH	7.7

Recommended Dilution:
    Winter use: 1 part cleaner with 1 part water.
    Summer use: May dilute to 1 part cleaner with 5 parts water.

SOURCE: Shell Chemical Co.: NEODOL Formulary: Formulations

WINDSHIELD WASHER FORMULATIONS

For Summer Use

RAW MATERIALS	% By Weight
1.	
DOWANOL PM glycol ether	16
DOWANOL DPM glycol ether	6
Pluronic L-62	.02
water	77.98

Dilute 1:1 with water for use concentration.

II.	
DOWANOL DPM glycol ether	5
DOWANOL PM glycol ether	5
Colloidal Silica	4
DOWFAX 2A-1 surfactant	.05
water	85.95

Use as is.

III.	
DOWANOL DPM glycol ether	6
DOWANOL PM glycol ether	16
isopropanol	10
Triton N 101	1
water	67

Dilute 1:1 with water for use concentration.

For Winter Use

IV.	
DOWANOL PM glycol ether	5
isopropanol	80
Ethylene Glycol	14
Igepal CO-630	1

Dilute 1:1 with water for use concentration.

V.	
DOWANOL PM glycol ether	12
Propylene Glycol	12
isopropanol	76

Dilute 1:1 with water for use concentration.

The winter formulations can be diluted up to 5 times for summer use.

SOURCE: Dow Chemical U.S.A.: The Glycol Ethers Handbook: Formulations

WINDSHIELD WASH

RAW MATERIALS	% By Weight
Methanol	40.0
Ammonium hydroxide (30% active)	1.0
EMULPHOGENE DA-630	4.0
Water	55.0
	100.0

Manufacturing Procedure:
Dissolve EMULPHOGENE DA-630 in water. Add methanol and
ammonium hydroxide. Mix thoroughly.

Physical Properties:
pH (as is)	10.3
pH (1%)	8.6
Viscosity	10 cps
Specific Gravity	.99

SOURCE: GAF Corp.: Formulary: Prototype Formulation GAF 5475

WINDSHIELD WASHER FORMULATION

RAW MATERIALS	% By Weight
SURFONIC N-95	0.2
Water	49.8
Metnanol or isopropanol	50.0
Alphazurine 2G Blue Dye	q.s.
Perfume	q.s.

Use Concentrations:
Winter: 8 ounces of concentrate plus 24 ounces water.
Summer: 2 ounces of concentrate plus 30 ounces water.

SOURCE: Texaco Chemical Co.: Formulation

Section III
Trademarked Raw Materials

RAW MATERIAL	CHEMICAL DESCRIPTION	SOURCE
ACRYSOL A-5 Thickener	Acrylic emulsion copolymer thickener/stabilizer	Rohm and Haas
ACRYSOL ASE-95 Thickener	High molecular weight acrylic emulsion copolymer(20% solids)	Rohm and Haas
ACRYSOL ASE-108 Thickener	High molecular weight acrylic thickener/stabilizer(20% solids)	Rohm and Haas
ACRYSOL ICS-1 Thickener	Alkali-soluble acrylic polymer emulsion(30% solids)	Rohm and Haas
ACRYSOL LMW-45 Polymer	Acrylic emulsion copolymer thickener	Rohm and Haas
ACTINOL FA-2 Tall Oil Fatty Acid	Tall oil fatty acid composition	Arizona Chemical
ACTRASOL SR606 Surfactant	Anionic surfactant. Oleic acid base. 35% free fatty acid	Arthur Trask
ADOGEN 470-75% Fabric Softener	Fabric softener	Sherex
AEROSIL 200 Fumed Silica	Fumed silica	Degussa
AEROSOL A-103 Surfactant	Disodium ethoxylated nonyl- phenol half ester of sulfo- succinic acid.	American Cyanamid
AEROSOL C-61 Surfactant	Alkylamine-guanidine poly- oxyethanol	American Cyanamid
AEROSOL MA-80 Surfactant	Sodium dihexyl sulfosuccinate. Anionic 80% liquid.	American Cyanamid
AEROSOL OT-S Surfactant	Sodium dioctyl sulfosuccinate. Anionic. 70%. Petroleum dist.	American Cyanamid
AEROSOL OT-B Surfactant	Sodium dioctyl sulfosuccinate. Anionic. 85% active powder.	American Cyanamid
AEROSOL OT-75% Surfactant	Sodium dioctyl sulfosuccinate. Anionic. 75% liquid.	American Cyanamid
AEROSOL 22 Surfactant	Tetrasodium N-(1,2-dicarboxy- ethyl)-N-octadecyl sulfo- succinate.	American Cyanamid

RAW MATERIAL	CHEMICAL DESCRIPTION	SOURCE
AEROTHENE MM Solvent	Specially inhibited grade of methylene chloride solvent	Dow Chemical
AEROTHENE TT Solvent	Aerosol grade 1,1,1-trichloro-ethane solvent	Dow Chemical
ALFOL 1214 Linear Alcohol	Primary linear alcohol	Vista Cnemical
ALIPAL CD-128 Surfactant	Surfactant	GAF
ALIPAL CO-433 Surfactant	Surfactant	GAF
ALIPAL CO-436 Surfactant	Surfactant	GAF
ALKAWET Surfactant	Industrial wetting agent	Lonza
ALOX 940 Surfactant	Surfactant	Alox
ALPHAZURINE 2G Blue Dye	Triphenylmethane acid blue.	Allied Chemical
AMMONYX LO Amine Oxide	Lauryl dimethylamine oxide	Stepan
AMPHOTERGE K Surfactant	Cocoampnopropionate. Amphoteric ionic character	Lonza
ANTAROX BL-225 Surfactant	Nonionic modified linear ali-phatic polyether	GAF
ANTAROX BL-240 Surfactant	Nonionic modified linear ali-phatic polyether	GAF
ANTAROX BL-330 Surfactant	Modifed linear aliphatic poly-ether(95% active)	GAF
ARMOSOFT WA104 Softener Base	Softener base	Akzo Chemie
AROMATIC 150 Solvent	Narrow-cut aromatic solvent	Exxon
AROSURF 42-PE 10 Surfactant	Alkylated tallow alcohol. 100% Conc.	Sherex

RAW MATERIALS	CHEMICAL DESCRIPTION	SOURCE
ATLAS G-3300 Emulsifier	Polyoxyethylene glyceride ester emulsifier	ICI Americas
AVANEL S-30 Surfactant	Sodium alkyl polyether sulfonate Molecular weight: 420	Mazer
AVANEL S-70 Surfactant	Sodium alkyl polyether sulfonate Molecular weight: 600	Mazer
BARDAC 22 (50%) Quaternary Compound	Quaternary ammonium sanitizing compound	Lonza
BARQUAT MD-50 (50%) Quaternary Compound	Quaternary ammonium sanitizing compound	Lonza
Berkeley 160 Mesh Supersil	Abrasive powder	U.S. Silica
Berkeley 230 Mesh (Jasper)	Abrasive powder	U.S. Silica
BIOPAL NR-20 Surfacant	Surfactant	GAF
BIOSOFT D-62 Surfactant	Sodium alkylbenzene sulfonate, linear. 60% active	Stepan
BIOSOFT EA-10 Surfactant	Fatty alcohol ethoxylate, mod- ified. 100% active	Stepan
BIOSOFT LD-150 Surfactant	Formulated detergent base. 50% active	Stepan
BIOSOFT LD-190 Surfactant	Formulated detergent base. 91% active	Stepan
BRITESIL C-24 Sodium Silicate	Soluble, clarified sodium silicate	PQ
BRITESIL H-20 Sodium Silicate	Soluble, clarified sodium silicate	PQ
BRITESIL 20 Sodium Silicate	Soluble, clarified sodium silicate	PQ

RAW MATERIALS	CHEMICAL DESCRIPTION	SOURCE
BUTYL CARBITOL Glycol Ether	Diethylene glycol monobutyl ether solvent	Union Carbide
BUTYL CELLOSOLVE Glycol Ether	Ethylene glycol monobutyl ether solvent	Union Carbide
BUTYL OXITOL Glycol Ether	Ethylene glycol monobutyl ether solvent	Shell Chemical
CALCOZINE Rhodamine BX Conc. Dye	Basic dye	American Cyanamid
CALIBRITE SL	Calcium carbonate	Hoechst
CALSOFT T-60 Sulfonate	Liquid trithanolamine dodecyl-benzene sulfonate(60% active)	Pilot
CARNAUBA SPRAY 200	Concentrated formulation base	Sherex
CARSPRAY CW Surfactant	Cationic/nonionic surfactant. Formulated as automobile product	Sherex
CARSPRAY Oil	Proprietary oil	Ashland
CARSPRAY #2 Concentrate	Concentrated formulation base. Blend of emulsifiers and solvents	Sherex
CARSPRAY 300 Emulsifier	Cationic emulsifier used as a base	Sherex
CARSPRAY 700 Carnauba Foamer	Concentrated formulation base containing modified Carnauba	Sherex
CDB CLEARON	Sodium dichloro-s-triazinetrione	Olin
CEDAMIDE AX Diethanolamide	Lauric diethanolamide	Miranol
CEDEMIDE CX Diethanolamide	Coco diethanolamide	Miranol
CEDEPAL SN-303 POE Sulfate	Sodium lauryloxy POE (2.0) sulfate	Miranol
CEDEPHOS FA 600M Phosphate Ester	Phosphate ester. 100% solids	Miranol

RAW MATERIAL	CHEMICAL DESCRIPTION	SOURCE
CELITE Diatomaceous Earth	Diatomaceous silica low cost pigment and flatting agent	Manville
CHEELOX BF-13 Sequestrant	EDTA tetrasodium salt. 89% active	GAF
CHEELOX BF-78 Sequestrant	Sequestrant	GAF
CHEELOX NTA-NA3 Sequestrant	Sequestrant	GAF
COBRATEC-99 Inhibitor	Benzotriazole(technical). Corrosion inhibitor	PMC Special.
CONOCO Sulfate A	Ammonium lauryl sulfate	Cont Chemical
CYCLO SOL 53 Solvent	Aromatic hydrocarbon, bp 325-349F	Shell Chemical
CYCLO SOL 63 Solvent	Aromatic solvent	Shell Chemical
DESODET 1239 Detergent	Detergent Blend	DeSoto
DESONAL SE Surfactant	Sodium Laureth Sulfate. 58-60% active	DeSoto
DESONATE AOS Surfactant	Sodium Alpha Olefin Sulfonate	DeSoto
DESONATE 60-S Surfactant	Sodium Linear Alkyl Benzene Sulfonate. 60% active	DeSoto
DESONIC 9N Surfactant	Nonyl Phenol Ethoxylate. 100% active	DeSoto
DESONIC 315-3 Surfactant	Surfactant	DeSoto
DESONIC 315-7 Surfactant	Surfactant	DeSoto

RAW MATERIAL	CHEMICAL DESCRIPTION	SOURCE
DESONIC 1036 Surfactant	Surfactant	DeSoto
DESONOL SE SLS	Sodium laureth sulfate	DeSoto
DESPHOS 5AP Phosphate Ester	Phosphate ester. 100% active	DeSoto
DIACID H-240 Surfactant	Previously neutralized potassium salt in water	Westvaco
DOWANOL DB Glycol Ether	Diethylene Glycol Butyl Ether	Dow Chemical
DOWANOL DE Glycol Ether	Diethylene Glycol Ethyl Ether	Dow Chemical
DOWANOL DPM Glycol Ether	Dipropylene Glycol Methyl Etner	Dow Chemical
DOWANOL EB Glycol Ether	Ethylene Glycol Butyl Ether	Dow Chemical
DOWANOL EE Glycol Ether	Ethylene Glycol Ethyl Ether	Dow Chemical
DOWANOL EM Glycol Ether	Ethylene Glycol Methyl Ether	Dow Chemical
DOWANOL PM Glycol Ether	Propylene Glycol Methyl Ether	Dow Chemical
DOWANOL P-Mix Glycol Ether	Propylene Glycol Methyl Ether Homologs	Dow Chemical
DOWANOL TPM Glycol Ether	Tripropylene Glycol Methyl Ether	Dow Chemical
DOWCLENE EC Solvent	Colorless liquid solvent. Boiling point 77-122C.	Dow Chemical
DOWFAX 2A-1 Surfactant	Anionic surface active agent	Dow Chemical
DUPONOL C Surfactant	Surfactant based on lauryl sulfate	DuPont
DURCAL 40	Calcium carbonate	Hoechst

RAW MATERIAL	CHEMICAL DESCRIPTION	SOURCE
EMEREST 2350 Glycol Stearate	Ethylene glycol monostearate	Emery
EMEREST 2665 PEG-600 Dioleate	More hydrophilic than EMEREST 2648. Liquid pour pt.: 19	Emery
EMERSAL 6400 Sodium Lauryl Sulfate	Versatile detergent. Liquid pour pt.: 15	Emery
EMERSAL 6453 Sodium Laureth-3 Sulfate	Detergent or foaming agent. Liquid pour pt.: <0	Emery
EMERSOL 211 Oleic Acid	Oleic acid	Emery
EMERY 6705 Phenoxy-ethanol	Solvent. Liquid pour pt.: 13	Emery
EMID 6500 Cocamide MEA Amide	Coconut monoethanolamide. Solid M.P.: 72	Emery
EMID 6514 Coconut Super Diethanolamide	Foam stabilizer, thickener and detergent component	Emery
EMID 6515 Coconut Super Diethanolamide	Boosts foam. Inhibits redeposition of soils	Emery
EMID 6533 Modified Coconut Diethanolamide	Emulsifier, thickening agent and moderate foamer	Emery
EMID 6538 Modified Coconut Diethanolamide	Good thickening and foam stabilizing properties	Emery
EMULPHOGENE BC-720 Emulsifier	Nonionic emulsifier	GAF
EMULPHOGENE BC-840 Emulsifier	Nonionic emulsifier	GAF
EMULPHOGENE DA-630 Emulsifier	Nonionic emulsifier	GAF
ESI-TERGE B-15 Surfactant	Amine condensate. Amine type 2-1. Nonionic	Emulsion Systems
ESI-TERGE DDBSA	Dodecylbenzene sulfonic acid	Emulsion Systems

RAW MATERIAL	CHEMICAL DESCRIPTION	SOURCE
ESI-TERGE HA-20 Surfactant	Modified amine condensate. Nonionic-anionic	Emulsion Systems
ESI-TERGE N-100 Surfactant	Nonionic surfactant. Polyethylene glycol ether type	Emulsion Systems
ESI-TERGE RT-61 Surfactant	Specially formulated. 90% active	Emulsion Systems
ESI-TERGE S-10 Surfactant	Amine condensate. Non-ionic. Amine Type: 1-1	Emulsion Systems
ESI-TERGE SXS Surfactant	Sodium xylene sulfonate	Emulsion Systems
ESI-TERGE T-60 Surfactant	Anionic detergent type. 60% solids	Emulsion Systems
ESI-TERGE 320 Surfactant	Phosphated nonylphenoxy polyethoxy ethanol. Anionic	Emulsion Systems
ESI-TERGE 330 Surfactant	Phosphated glycol ester. 99% active. Anionic	Emulsion Systems
ETHOXYLAN 1686 PEG-75 Lanolin	Etnoxylated lanolin	Emery
FLUORAD FC-129 Surfactant	Liquid fluorochemical surfactant	3M
GAFAC LO-529 Surfactant	Sodium salt of complex organic phosphate ester. Anionic.	GAF
GAFAC PE-510 Surfactant	Complex organic phosphate ester. Anionic	GAF
GAFAC RA-600 Surfactant	Complex organic phosphate ester. 100% active	GAF
GAFAC RE-610 Surfactant	Complex organic phosphate ester. Anionic	GAF
GAFAMIDE CDD-518 Ethanolamide	Coconut oil diethanolamine condensate (100% active)	GAF

RAW MATERIALS	CHEMICAL DESCRIPTION	SOURCE
GANTREZ AN-149 Polymer	Vinyl Ethyl-Maleic Anhydride Copolymer	GAF
GENAMINOX CS Amine Oxide	Amine oxide, 30%	Hoechst/ Celanese
GENAMINOX KC Surfactant	Alkyldimethyl amine oxide. Active: 30%	Hoechst/ Celanese
GENAPOL PF 20 Condensate	Polyoxipropylen-polyoxiethylen Condensate	Hoechst/ Celanese
HALANE Compound	Dichloromethylhydantoin.	BASF
HAMPENE 100 Chelating Agent	Tetrasodium ethylenediamine-tetraacetate solution. 38% active	Hampshire
HAMPOSYL L-30 Surfactant	Sodium Lauroyl Sarcosinate. Active: 30%	Hampshire
HOE S 2817	Hydrotrope	Hoechst/ Celanese
HOECHST-WACHS VP KST	Wax	Hoechst/ Celanese
HOSTAPUR SAS 30 Detergent Base	N-alkane sulphonates. Active: 30%	Hoechst/ Celanese
HOSTAPUR SAS 60 Detergent Base	N-alkane sulphonates. Active: 60%	Hoechst/ Celanese
HYAMINE 3500 Germicide	Germicide	Rohm and Haas
ICONOL DA-6 Surfactant	POE (6) decyl alcohol surfactant	BASF
ICONOL TDA Surfactant	POE tridecyl alcohol	BASF

RAW MATERIAL	CHEMICAL DESCRIPTION	SOURCE
ICONOL TDA-8 Surfactant	POE (8) tridecyl alcohol	BASF
ICONOL TDA-10 Surfactant	POE (10) tridecyl alcohol	BASF
IGEPAL CA-620 Surfactant	Nonylphenoxy poly(ethyleneoxy) ethanol. Mole ratio: 7	GAF
IGEPAL CO Surfactant	Nonylphenoxy poly(ethyleneoxy) ethanol.	GAF
IGEPAL CO-430 Surfactant	Nonylphenoxy poly(ethyleneoxy) ethanol. Mole ratio: 4	GAF
IGEPAL CO-530 Surfactant	Nonylphenoxy poly(ethyleneoxy) ethanol. Mole ratio: 6	GAF
IGEPAL CO-620 Surfactant	Nonylphenoxy poly(ethyleneoxy) ethanol. Mole ratio: 8.5	GAF
IGEPAL CO-630 Surfactant	Nonylphenoxy poly(ethyleneoxy) ethanol. Mole ratio: 9	GAF
IGEPAL CO-660 Surfactant	Nonylphenoxy poly(ethyleneoxy) ethanol. Mole ratio: 10	GAF
IGEPAL CO-710 Surfactant	Nonylphenoxy poly(ethyleneoxy) ethanol. Mole ratio: 10-11	GAF
IGEPAL CO-730 Surfactant	Nonylphenoxy poly(ethyleneoxy) ethanol. Mole ratio: 15	GAF
IGEPAL CO-970 Surfactant	Nonylphenoxy poly(ethyeleneoxy) ethanol. Mole ratio: 50	GAF
IGEPAL RC-520 Surfactant	Surfactant	GAF
IGEPON TC-42 Surfactant	N-coconut acid-N-methyl-taurate (24% active)	GAF
INDUSTROL N3 Surfactant	Nonionic surfactant	BASF
IRGASAN DP 300 Bacteriostat	Bacteriostat	Ciba- Geigy

RAW MATERIALS	CHEMICAL DESCRIPTION	SOURCE
KAOPOLITE Aluminum Silicate	Anhydrous aluminum silicate	Georgia Kaolin
KAOPOLITE SF Aluminum Silicate	Anhydrous aluminum silicate	Georgia Kaolin
KASIL #1 Potassium Silicate	Liquid potassium silicate. Weight ratio: 2.50	PQ
KASIL #6 Potassium Silicate	Clear potassium silicate. Weight ratio: 2.10	PQ
KELZAN Thickener	Xanthan gum	Kelco
KLEARFAC AA-270 Surfactant	Phosphate ester surfactant. 90% active	BASF
Latex E-284(40%)	Latex	Morton
d-Limonene Oil	Oil derived from citrus pulps and peels	Sunkist/ or Union Camp
LOSER GX 5	Alkylarylpolyglycolether	Hoechst/ Celanese
LUDOX Tech. Colloidal Silica	Aqueous colloidal silica sol	DuPont
MAKON NF-5 Surfactant	Polyalkoxylated aliphatic base. 97% active	Stepan
MAKON 4 Surfactant	Nonoxynol-4 surfactant	Stepan
MAKON 10 Surfactant	Nonoxynol-10 surfactant	Stepan
MAKON 12 Surfactant	Nonoxynol-12 surfactant	Stepan
MAZER MACOL NP 9.5 Surfactant	Nonylphenol. Nonionic-type. HLB Value: 12.9	Mazer
MAZER MACOL OP-10 SP Surfactant	Octylphenol. Nonionic-type. HLB Value: 13.4	Mazer

RAW MATERIALS	CHEMICAL DESCRIPTION	SOURCE
MAZER MACOL 19 Surfactant	Block polyol. Nonionic. Molecular weight 2,200	Mazer
MAZER MACOL 24 Rinse Aid	100% active biodegradable. Molecular weight 800	Mazer
MAZER MACOL 25 Rinse Aid	100% active biodegradable. Molecular weight 1,000	Mazer
MAZER MACOL 30 Rinse Aid	100% active biodegradable. Molecular weight 600	Mazer
MAZER MACOL 40 Block Polyol	Nonionic. Molecular Weight 3,100	Mazer
MAZER MACOL 41 Surfactant	Ammonium salt of an alkylphenol ethoxylate	Mazer
MAZER MACOL 45 Rinse Aid	100% active biodegradable. Molecular weight 1,100	Mazer
MAZER MACOL 48 Surfactant	Polyoxyxethylene fatty ether	Mazer
MAZER MACOL 212 Surfactant	Ethoxylated fatty alcohol non- ionic surfactant	Mazer
MAZER MAFO 13 Surfactant	Amphoteric. Potassium salt of a complex amine carboxylate. 70%	Mazer
MAZER MAPHOS 60A Phosphate Ester	Complex phosphorylated nonionic. Aliphatic. 99.5 active	Mazer
MAZER MAPHOS 66H Phosphate Ester	Complex phosphorylated nonionic. Aromatic. 50% active	Mazer
MAZER MAPHOS 76NA Phosphate Ester	Complex phosphorylated nonionic. Aromatic. 99.5% active	Mazer
MAZER MAPHOS 91 Phosphate Ester	Complex phosphorylated nonionic. Aromatic. 99.5% active	Mazer
MAZER MASIL 1066C Silicone Glycol	Organo-modified silicone fluid. Alkylene oxide modified. 1800 cst.	Mazer
MAZER MAZAMIDE 80 Alkanolamide	Alkanolamide. 1:1 type. Coconut fatty acid. Nonionic	Mazer

RAW MATERIAL	CHEMICAL DESCRIPTION	SOURCE
MAZER MAZAWET DF Surfactant	Wetting agent. Nonionic. 100% active	Mazer
MAZER MAZAWET DOSS (70%) Surfactant	Wetting agent. Anionic. 70% active	Mazer
MAZER MAZAWET 77 Surfactant	Wetting agent. Nonionic. 100% active	Mazer
MAZER MAZON DDBSA	Dodecylbenzenesulfonic acid	Mazer
MAZER MAZON 40 Surfactant	Caustic coupling surfactant. Activity 70	Mazer
MAZER MAZON 41 Surfactant	Ammonium salt of an alkylphenol-ethoxylate. Activity 60	Mazer
MAZER MAZON 60T Surfactant	Alkylaryl sulfonate triethanol amine. Activity 60	Mazer
MAZER MAZON 71 Surfactant	Surfactant	Mazer
MAZER MAZON 71A Surfactant	Surfactant	Mazer
Medialan LD Anionic Surfactant	Lauroyl sarcoside sodium salt. Active: 30%	Hoechst/ Celanese
Methocel E4M Premium Thickener	Methylcellulose protective colloid, thickener	Dow Chemical
Methocel 65HG 4000 Thickener	Methylcellulose protective colloid, thickener	Dow Chemical
Miramine OC Surfactant	Surface active agent	Miranol
Miranol CM-SF Conc. Surfactant	Cocoamphopropionate. 37.0% solids	Miranol
Miranol C2M Conc. N.P. Surfactant	Cocoamphocarboxyglycinate. 50.0% solids	Miranol
Miranol C2M-SF Conc. Surfactant	Cocoamphocarboxyglycinate. 39.0% solids	Miranol

RAW MATERIAL	CHEMICAL DESCRIPTION	SOURCE
MIRANOL C2M-SF(70%) Surfactant	Cocoamphocarboxypropionate amphoteric (70%)	Miranol
MIRANOL CS Conc. Surfactant	Cocoamphopropylsulfonate. Amphoteric. 45.0% solids	Miranol
MIRANOL DM Surfactant	Stearoamphoglycinate. Amphoteric. 26.0% solids	Miranol
MIRANOL H2M Conc. Surfactant	Lauroamphocarboxypropionate. Amphoteric. 39.0% solids	Miranol
MIRANOL JEM Conc. Surfactant	Mixed C8 amphocarboxylates. Amphoteric. 34.0% solids	Miranol
MIRANOL J2M Conc. Surfactant	Caprylamphocarboxyglycinate. Amphoteric. 49.0% solids	Miranol
MIRANOL J2M-SF Conc. Surfactant	Caprylamphocarboxypropionate. Amphoteric. 38.5% solids	Miranol
MIRANOL JS Conc. Surfactant	Caprylamphocarboxypropylsulfonate Amphoteric. 49.0% solids	Miranol
MIRAPON JAS-50 Surfactant	Capryloamphopropionate. 50.0% Industrial. 50.0% solids	Miranol
MIRATAINE CBC Surfactant	Cocamidopropyl Betaine. Amphoteric. 35.0% solids	Miranol
MIRATAINE H2C Surfactant	Disodium Lauriminodipropionate. Amphoteric. 30.0% solids	Miranol
MIRAWET ASC Wetting Agent	Alkylether Hydroxypropyl Sultaine. 50.0% solids	Miranol
MIRAWET B Wetting Agent	Sodium Butoxyethoxy Acetate. 46.0% solids	Miranol
MIRAWET FL Wetting Agent	Modified amphoteric	Miranol
MONAWET MM-80 Surfactant	Sodium dialkylsulfosuccinate. 80% activity	Miranol
M-PYROL Solvent	N-methyl-2-pyrrolidone solvent	GAF

RAW MATERIAL	CHEMICAL DESCRIPTION	SOURCE
NEKAL WT-27 Surfactant	Surfactant	GAF
NEOCRYL A550 Acrylic	Acrylic polymer	Polyvinyl
NEODOL 15-S-9 Ethoxylate	Linear primary alcohol	Shell
NEODOL 23-3 Ethoxylate	Linear primary alcohol MW: 322. 39.6% EO	Shell Chemical
NEODOL 23-5 Ethoxylate	Linear primary alcohol	Shell Chemical
NEODOL 23-6.5 Ethoxylate	Linear primary alcohol. MW: 488. 60.4% EO	Shell Chemical
NEODOL 25-3 Ethoxylate	Linear primary alcohol. MW: 338. 39.0% EO	Shell Chemical
NEODOL 25-3A Ethoxysulfate	Ethoxysulfate. Ammonium cation. Active: 59%	Shell Chemical
NEODOL 25-3S Ethoxysulfate	Ethoxysulfate. Sodium cation. Active: 59%	Shell Chemical
NEODOL 25-7 Ethoxylate	Linear primary alcohol. MW: 524. 61.3% EO	Shell Chemical
NEODOL 25-9 Ethoxylate	Linear primary alcohol. MW: 610. 65.6% EO	Shell Chemical
NEODOL 25-12 Ethoxylate	Linear primary alcohol. MW: 729. 71.8% EO	Shell Chemical
NEODOL 45-2.25 Ethoxylate	Linear primary alcohol. MW: 319. 31.6% EO	Shell Chemical
NEODOL 45-7 Ethoxylate	Linear primary alcohol. MW: 529. 59.0% EO	Shell Chemical
NEODOL 91-2.5 Ethoxylate	Linear primary alcohol. MW: 281. 42.3% EO	Shell Chemical
NEODOL 91-6 Ethoxylate	Linear primary alcohol. MW: 428. 62.7% EO	Shell Chemical
NEODOL 91-8 Ethoxylate	Linear primary alcohol. MW: 519. 69.5% EO	Shell Chemical

RAW MATERIAL	CHEMICAL DESCRIPTION	SOURCE
NINATE 411 Surfactant	Alkylamine dodecylbenzene sulfonate. 93% active	Stepan
NINOL 11-CM Surfactant	Coconut diethanolamide	Stepan
NINOL 49CE Surfactant	Fatty acid diethanolamide	Stepan
NINOL 128-Extra Surfactant	Surfactant	Stepan
NINOL 2012EX Surfactant	Fatty acid diethanolamide	Stepan
NINOL 1281 Surfactant	Fatty acid alkylolamide. 100% active	Stepan
NINOL 1285 Surfactant	Fatty acid base alkylolamide. 100% acid	Stepan
NOPCOSTAT HS Lubricant	Antistatic lubricant	Diamond Shamrock
ORVUS K Surfactant	Liquid surfactant	Procter & Gamble
PETRO BA Liquid Surfatrope	Alkyl Naphthalene Sodium Sulfonate. 50% active	DeSoto
PETRO BA Powder Surfatrope	Alkyl Naphthalane Sodium Sulfonate. 95% active	DeSoto
PETRO BAF Liquid Surfatrope	Alkyl Naphthalene Sodium Sulfonate. 50% active	DeSoto
PETRO BAF Powder Surfatrope	Alkyl Naphthalene Sodium Sulfonate. 95% active	DeSoto
PETRO LBA Liquid Surfatrope	Alkyl Naphthalene Sodium Sulfonate. 50% active	DeSoto
PETRO LBA Powder Surfatrope	Alkyl Naphthalene Sodium Sulfonate. 95% active	DeSoto

RAW MATERIALS	CHEMICAL DESCRIPTION	SOURCE
PETRO ULF Surfatrope	Alkyl naphthalene sodium sulfon- ate. 50% active. Liquid	DeSoto
PETRO 22 Surfatrope	Alkyl naphthalene sodium sulfon- ate. 50% active. Liquid	DeSoto
PHOSPHAT SPR II	Proprietary	Hoechst
PLURAFAC A-38 Surfactant	Linear alcohol alkoxylate. HLB: 20	BASF
PLURAFAC B-25-5 Surfactant	Linear alcohol alkoxylate. HLB: 12	BASF
PLURAFAC B-26 Surfactant	Linear alcohol alkoxylate. HLB: 14	BASF
PLURAFAC C-17 Surfactant	Linear alcohol alkoxylate. HLB: 16	BASF
PLURAFAC D-25 Surfactant	Linear alcohol alkoxylate. HLB: 10	BASF
PLURAFAC RA-20 Surfactant	Linear alcohol alkoxylate. HLB: 10	BASF
PLURAFAC RA-30 Surfactant	Linear alcohol alkoxylate. HLB: 9	BASF
PLURAFAC RA-40 Surfactant	Linear alcohol alkoxylate. HLB: 7	BASF
PLURAFAC RA-43 Surfactant	Straight chain alcohol. HLB: 7	BASF
PLURONIC F108 Surfactant	Liquid polyol. MW: 14000. Viscosity: 8,000 cps.	BASF
PLURONIC L10 Surfactant	Liquid polyol. HLB: 12-18	BASF
PLURONIC L61 Surfactant	Liquid polyol. HLB: 1-7	BASF
PLURONIC L62 Surfactant	Liquid polyol. MW: 2,500. HLB: 1-7	BASF
PLURONIC L62D Surfactant	Liquid polyol. MW: 2,500. HLB: 1-7	BASF

RAW MATERIALS	CHEMICAL DESCRIPTION	SOURCE
PLURONIC 25R2 Surfactant	Block copolymer nonionic. MW: 3,100	BASF
POLECTRON 430 Copolymer	Vinylpyrrolidone copolymer	GAF
PRIMAPEL C-93 Polymer(25%)	Water-borne carboxylated acrylic copolymer(25% active)	Rohm and Haas
PROPASOL BEP Solvent	Glycol ether	Union Carbide
RENEX 36 Surfactant	Polyoxyethylene (6) tridecyl ether surfactant	ICI Americas
REWOCOR B3010 Surfactant	Surfactant	Sherex (Rewo)
REWOPOL FBR Surfactant	Nonionic surfactant	Sherex (Rewo)
REWOPOL HV9 Surfactant	Nonionic surfactant	Sherex (Rewo)
REWOPOL HV10 Surfactant	Nonionic surfactant	Sherex (Rewo)
REWOPOL PCK 2000 Surfactant	Nonionic surfactant	Sherex (Rewo)
REWOPOL PEG 6000 DS Surfactant	PEG 150 Distearate	Sherex (Rewo)
REWOQUAT B50 Surfactant	Imidazoline	Sherex (Rewo)
REWORYL NXS 40 Surfactant	Na. xylene sulfonate 40%	Sherex (Rewo)
REWORYL NXS 50 Surfactant	Surfactant	Sherex (Rewo)
REWORYL TXS 90/F Surfactant	Surfactant	Sherex (Rewo)
REWOTERIC AM-V Surfactant	Quaternary disinfectant and sanitizer	Sherex (Rewo)

RAW MATERIAL	CHEMICAL DESCRIPTION	SOURCE
SHELL Mineral Spirits 145,150 or 150 EC	Mineral spirits	Shell Chemical
SHELL SOL 72 Solvent	Isoparaffinic solvent, bp 356-401F	Shell Chemical
SHELL SOL 140 Solvent	Solvent	Shell Chemical
SHELL SOL 340 Solvent	Aliphatic naphtha, flash point 104F, bp 316-358F	Shell Chemical
SIPON L-22(28%) Lauryl Sulfate	Ammonium lauryl sulfate	Alcolac
SIPON LSB Lauryl Sulfate	Sodium lauryl sulfate(29% active)	Alcolac
SIPONATE A-246L Alkyl Aryl Sulfonate	Alkyl aryl sulfonate	Alcolac
SMA 1000A Powder Copolymer	Styrene maleic anhydride copolymer	Arco
SMA 2625 Resin Copolymer	Styrene-maleic anhydride copolymer. MW: 1,900	Arco
SPAN 80 Surfactant	Sorbitan monooleate	ICI Americas
STARSO Sodium Silicate	Liquid sodium silicate. Weight ratio: 1.80	PQ
STEOL CS-460 Surfactant	Fatty ether sulfate. Sodium cation. 60% active	Stepan
STEPANATE X Surfactant	Xylene hydrotrope. Sodium cation. 40% active	Stepan
STEPANOL WA-Special Surfactant	Fatty alcohol alkyl sulfate. Sodium cation. 29% active	Stepan
STEPANOL WAC Surfactant	Sodium lauryl sulfate	Stepan
SUPER FLOSS Silica	Processed diatomaceous silica	Manville

RAW MATERIAL	CHEMICAL DESCRIPTION	SOURCE
SURFONIC HDL Surfactant	Surface-active agent	Texaco
SURFONIC JL-80X Surfactant	Alkylpolyalkoxyethanol. Nonionic	Texaco
SURFONIC LF-17 Surfactant	Nonionic ethylene oxide adduct	Texaco
SURFONIC N-40 Surfactant	Nonionic surface-active agent. HLB: 8.9. 100% active	Texaco
SURFONIC N-60 Surfactant	Nonionic surface-active agent. HLB: 10.9. 100% active	Texaco
SURFONIC N-85 Surfactant	Nonionic surface-active agent. HLB: 12.6. 100% active	Texaco
SURFONIC N-95 Surfactant	Nonionic surface-active agent. HLB: 12.9. 100% active	Texaco
SURFONIC N-100 Surfactant	Nonionic surface-active agent. HLB: 13.2. 100% active	Texaco
SURFONIC N-102 Surfactant	Nonionic surface-active agent. HLB: 13.4. 100% active	Texaco
SURFONIC N-120 Surfactant	Nonionic surface-active agent. HLB: 14.1. 100% active	Texaco
SURFONIC N-150 Surfactant	Nonionic surface-active agent. HLB: 15.0. 100% active	Texaco
TAMOL SN Dispersant	Anionic polymer-type dispersing agent	Rohm and Haas
TEGO Betaine L-7	Betaine	Hoechst/ Celanese
TERGESCENT No. 7	Proprietary	Givaudan
THERMPHOS NW	Proprietary	Hoechst/ Celanese

RAW MATERIALS	CHEMICAL DESCRIPTION	SOURCE
TINOPAL CBS-X Whitening Agent	Distyryl biphenyl disulfonate derivative. Fluorescent	Ciba- Geigy
TINOPAL RBS-2000 Whitening Agent	Naphthotriazolstilbene monosul- fonate derivative. Fluroescent	Ciba- Geigy
TINOPAL UNPA Whitening Agent	Whitening agent	Ciba- Geigy
TINOPAL 5BM Whitening Agent	Whitening agent	Ciba- Geigy
TRILON B, Liquid	EDTA Sodium Salt Solution	BASF
TRITON BG-10 Surfactant	Biodegradable, nonionic. 70% active	Rohm and Haas
TRITON CF-10 Surfactant	Alkylaryl polyether nonionic. 100% active	Rohm and Haas
TRITON CF-21 Surfactant	Alkylaryl polyether nonionic. 100% active	Rohm and Haas
TRITON CF-32 Surfactant	Amine polyglycol condensate. 100% active	Rohm and Haas
TRITON CF-54 Surfactant	Modified polyethoxy adduct. 100% active	Rohm and Haas
TRITON CF-76 Surfactant	Modified polyethoxy adduct non- ionic. 100% active	Rohm and Haas
TRITON CF-87 Surfactant	Modified polyethoxy adduct. 90% active	Rohm and Haas
TRITON DF-12 Surfactant	Modified polyoxyethylated alcohol nonionic. 100% active	Rohm and Haas
TRITON DF-16 Surfactant	Terminated ethoxylated linear alcohol. 100% active	Rohm and Haas
TRITON DF-18 Surfactant	Biodegradable modified alcohol nonionic. 90% active	Rohm and Haas
TRITON DF-20 Surfactant	Modified ethoxylate, acid form. 100% active	Rohm and Haas

RAW MATERIAL	CHEMICAL DESCRIPTION	SOURCE
TRITON GR-7M Surfactant	Dioctyl sodium sulfosuccinate. 64% active	Rohm and Haas
TRITON H-55 Surfactant	Phosphate ester, salt form 50% active	Rohm and Haas
TRITON H-66 Surfactant	Phosphate ester, salt form anionic. 50% active	Rohm and Haas
TRITON N-42 Surfactant	Surfactant	Rohm and Haas
TRITON N-57 Surfactant	Nonylphenol	Rohm and Haas
TRITON N-87 Surfactant	Nonylphenol	Rohm and Haas
TRITON N-101 Surfactant	Nonylphenol. 9-10 mols EO. Nonionic. 100% active	Rohm and Haas
TRITON N-998 Surfactant	Nonylphenol	Rohm and Haas
TRITON QS-15 Surfactant	Sodium salt of amphoteric. 100% active	Rohm and Haas
TRITON QS-30 Surfactant	Phosphate ester, acid form, anionic. 90% active	Rohm and Haas
TRITON QS-44 Surfactant	Phosphate ester, acid form, anionic. 80% active	Rohm and Haas
TRITON RW Surfactant	Surfactant	Rohm and Haas
TRITON X-45 Surfactant	Octylphenol. 5 mols EO. Nonionic. 100% active	Rohm and Haas
TRITON X-55 Surfactant	Octylphenol	Rohm and Haas
TRITON X-100 Surfactant	Octylphenol. 9-10 mols EO. 100% active	Rohm and Haas
TRITON X-102 Surfactant	Octylphenol. 12-13 mols EO. 100% active	Rohm and Haas

RAW MATERIAL	CHEMICAL DESCRIPTION	SOURCE
TRITON X-114 Surfactant	Octylphenol. 7-8 mols EO. 100% active	Rohm and Haas
TRITON X-207 Surfactant	Octylphenol	Rohm and Haas
TRITON X-301 Surfactant	Sodium alkylaryl polyether sulfate. Anionic. 20% active	Rohm and Haas
TRYCOL 5940 POE(6) Tridecyl Alcohol	Ethoxylated alcohol. HLB: 11.4	Emery
TRYCOL 5941 POE(9) Tridecyl Alcohol	Ethoxylated alcohol. Nonionic. HLB: 13.0	Emery
TRYCOL 5943 POE(12) Tridecyl Alcohol	Ethoxylated alcohol. Hydrophilic. HLB: 14.5	Emery
TRYCOL 5951 POE(5) Decyl Alcohol	Ethoxylated alcohol	Emery
TRYCOL 5966 Ethoxylated Alcohol	Ethoxylated alcohol. HLB: 8.7	Emery
TRYCOL 5967 POE(12) Lauryl Alcohol	Ethoxylated alcohol. HLB: 14.4	Emery
TRYCOL 6952 POE(15) Nonylphenol	Ethoxylated alkylphenol. HLB: 15.0	Emery
TRYCOL 6953 POE(12) Nonylphenol	Ethoxylated alkylphenol. HLB: 14.1	Emery
TRYCOL 6961 POE(4) Nonylphenol	Ethoxylated alkylphenol. HLB: 8.9	Emery
TRYCOL 6964 POE(9) Nonylphenol	Ethoxylated alkylphenol. HLB: 13.0	Emery
TRYCOL 6965 POE(11) Nonylphenol	Ethoxylated alkylphenol. HLB: 13.5	Emery
TRYFAC 5552 Phosphate Ester	Free acid form. Anionic. Pour Pt.: <-15.	Emery
TRYFAC 5553 Phosphate Ester	Potassium salt of TRYFAC 5552. Pour Pt.: <-15.	Emery

RAW MATERIALS	CHEMICAL DESCRIPTION	SOURCE
TRYFAC 5555 Complex Phosphate Ester	Free acid form. Most hydrophobic ester of series. Pour Pt.: -3	Emery
TRYFAC 5556 Complex Phosphate Ester	Free acid form. Pour Pt.: 5	Emery
TRYFAC 5559 Phosphate Ester	Water soluble detergent. Pour Pt.: 18	Emery
TRYFAC 5568 HWD Hydrotrope	Phosphate ester in free acid form 100% active. Pour Pt.: 12	Emery
TRYFAC 5569 Phosphate Ester	Free acid form. Pour Pt.: 5	Emery
TRYFAC 5576 Phosphate Ester	Potassium salt. Pour Pt: <-10	Emery
TRYLON 6735 Nonionic Wetting Agent	Low-foaming emulsifier. Pour Pt.: 9	Emery
TRYMEEN 6606 POE(15) Tallow Amine	Ethoxylated fatty amine. HLB: 14.3	Emery
TWEEN 60 Surfactant	Polyoxyethylene (20) sorbitan monostearate	Emery
TYLOSE CBR 4000	Water-soluble cellulose ether	Hoechst
TYLOSE CBR 10 000	Water-soluble cellulose ether	Hoecsht
UDET 950 Surfactant	Linear alkyl aryl sulfonate. 95% active	DeSoto
ULTRAWET 45KX Surfactant	Linear alkylbenzene sulfonate	Rohm and Haas
VANGEL B Thixotrope	Natural smectite clay	RT Vanderbilt
VARAMIDE A-2 Surfactant	Coconut diethanolamide. 100% Conc.	Sherex

RAW MATERIAL	CHEMICAL DESCRIPTION	SOURCE
VARAMIDE A-10 Surfactant	Modified coco diethanolamide 100% conc.	Sherex
VARAMIDE FBR Surfactant	Surfactant blend. 100% conc.	Sherex
VARAMIDE MA-1 Surfactant	Alkanolamide	Sherex
VARAMIDE ML-1 Surfactant	Alkanolamide	Sherex
VARAMIDE 6CM Surfactant	Alkanolamide	Sherex
VARIDRI 40 Surfactant	Surfactant	Sherex
VARINE O Surfactant	Alkyl Hydroxy-ethyl-imid-azoline. Alkyl: Oleic	Sherex
VARION AMK-SF Surfactant	Salt free amphoteric. 40% Conc.	Sherex
VARION AM-V Surfactant	Amphoteric glycine derivative. 35% Conc.	Sherex
VARION CADG Surfactant	Cocoamidopropyl betaine. 35% Conc.	Sherex
VARION CDG Surfactant	Lauryl betaine. 35% Conc.	Sherex
VARION EP AMVSF Surfactant	Amphoteric surfactant	Sherex
VARION HC Surfactant	High alkaline stable amphoteric. 50% Conc.	Sherex
VARION TEG Surfactant	Tallow amphoteric. 48% Conc.	Sherex
VARION 2C Surfactant	Amphoteric glycine derivative. 50% Conc.	Sherex
VARIQUAT 50MC Surfactant	Benzalkonium chloride. 50% Conc.	Sherex

RAW MATERIAL	CHEMICAL DESCRIPTION	SOURCE
VARISOFT 222LT-90% Fabric Softener	Softener base	Sherex
VARISOFT 3690-75% Fabric Softener	Liquid imidazoline quaternary	Sherex
VARISOFT 3690-90% Fabric Softener	Liquid imidazoline quaternary	Sherex
VARISOFT 3690N-90% Fabric Softener	Liquid imidazoline quaternary	Sherex
VAROX 185E Surfactant	Alkyl ether amine oxide. 40% Conc.	Sherex
VAROX 365 Surfactant	Lauryl dimethyl amine oxide	Sherex
VAROX 1770 Surfactant	Cocamidopropyl amine oxide. 35% Conc.	Sherex
VARSULF S1333 Surfactant	Ricinoleic sulfosuccinate. 40% Conc.	Sherex
VARSULF SBL203 Surfactant	Fatty acid alkanolamide sulfo-succinate. 40% Conc.	Sherex
VEEGUM Stabilizer	Complex colloidal magnesium aluminum silicate	RT Vand-erbilt
VEEGUM T Stabilizer	Complex colloidal magnesium aluminum silicate	RT Vand-erbilt
VERSENE 100 Chelating Agent	Tetrasodium ethylene tetra-acetate	Dow Chemical
WESTVACO L-5	Fatty acid	Westvaco
WITCOLATE D-510 Surfactant	Surfactant	Witco
WITCONATE SXS 40 Surfactant	Sodium xylene sulfonate	Witco
WITCONATE 1238 LAS	C12 LAS	Witco
ZEOLITE 4A	Zeolite	PQ

Section IV
Suppliers' Addresses

Akzo Chemie
300 S. Wacker Dr.
Chicago, IL 60606

Alcolac, Inc.
3440 Fairfield Rd.
Baltimore, MD 21226

Alox Corp.
3943 Buffalo Ave.
Niagara Falls, NY 14302

American Cyanamid Co.
One Cyanamid Plaza
Wayne, NJ 07470

Arizona Chemical Co.
200 S. Suddeth Pl.
Panama City, FL 32404

Ashland Chemical Co.
P.O. Box 2219
Columbus, OH 43216

BASF Corp.
100 Cherry Hill Rd.
Parsippany, NJ 07054

Ciba-Geigy Corp.
Three Skyline Drive
Hawthorne, NY 10532

Continental Chemical Co.
270 Clifton Blvd.
Clifton, NJ 07015

Degussa Corp.
Rt. 46 at Hollister Rd.
Teterboro, NJ 07608

DeSoto, Inc.
1700 S. Mt. Prospect Rd.
Des Plaines, IL 60018

Diamond Shamrock Chemicals
351 Phelps Ct.
Irving, TX 75038

Dow Chemical U.S.A.
Midland, MI 48640

DuPont Co.
Market St.
Wilmington, DE 19898

Emery Chemicals
11501 Northlake Dr.
P.O. Box 429557
Cincinnati, OH 45249

Emulsion Systems, Inc.
215 Kent Ave.
Brooklyn, NY 11211

Exxon Chemical Americas
13501 Katy Frwy.
Houston, TX 77079

GAF Corp.
1361 Alps Rd.
Wayne, NJ 07470

Georgia Kaolin Co., Inc.
2700 US Highway 22 East
P. O. Box 3110
Union, NJ 07083

Givaudan Corp.
100 Delawanna Ave.
Clifton, NJ 07014

Hampshire Organic Chemicals Div.
W.R. Grace & Co.
Poisson Ave.
Nashua, NH 03061

Hoechst/Celanese
4331 Chesepeake Drive
P.O. Box 16267
Charlotte, NC 28216

ICI Americas, Inc.
Concord Pike & New Murphy Rd.
Wilmington, DE 19897

Kelco Division
Merck & Co., Inc.
20 North Wacker Drive
Chicago, IL 60606

Lonza, Inc.
22-10 Rte. 208
Fair Lawn, NJ 07410

Manville Filtration/Minerals
P.O. Box 5108
Denver, CO 80217

Mazer Chemicals, Inc.
3938 Porett Drive
Gurnee, IL 60031

Miranol Chemical Co., Inc.
68 Culver Rd.
P.O. Box 436
Dayton, NJ 08810

Morton Thiokol, Inc.
Morton Chemical Division
333 W. Wacker Drive
Chicago, IL 60606

Olin Corp.
120 Long Ridge Rd.
P.O. Box 1355
Stamford, CT 06904-1355

PMC Specialties Corp., Inc.
20525 Center Ridge Rd.
Rocky River, OH 44116

PQ Corp.
P.O. Box 840
Valley Forge, PA 19482

Pilot Chemical Co.
11756 Burke St.
Santa Fe Springs, CA 90670

Polyvinyl Chemicals, Inc.
730 Main St.
Wilimngton, MA 01887

Procter & Gamble Industries
Chemical Division
P.O. Box 599
Cincinnati, OH 45201

Rohm and Haas Co.
Independence Mall West
Philadelphia, PA 19105

Shell Chemical Co.
1 Shell Plaza
Houston, TX 77002

Sherex Chemical Co., Inc.
5777 Frantz Rd.
P.O. Box 646
Dublin, OH 43017

Stepan Co.
22 W. Frontage Rd.
Northfield, IL 60093

Sunkist Growers, Inc.
14130-T Riverside Drive
Sherman Oaks, CA 91423

3M Co.
3M Center
2501 Hudson Road
St. Paul, MN 55100

Texaco Chemical Co.
480 Fournace Rd.
P.O. Box 430
Bellaire, TX 77401

Arthur C. Trask Corp.
7666 West 63rd St.
Summit, IL 60501

Union Camp Corp.
Chemical Division
1600 Valley Rd.
Wayne, NJ 07470

Union Carbide Corp.
Old Ridgebury Rd.
Danbury, CT 06817

U.S. Silica
P.O. Box 187
Berkeley Springs, WV 25411

R.T. Vanderbilt Co., Inc.
30 Winfield St.
Norwalk, CT 06855

Vista Chemical
P.O. Box 19029
15990 Barkers Landing Rd.
Houston, TX 77224

Westvaco Corp.
Chemical Division
P.O. Box 70848
Cherleston Heights, SC 29415

Witco Corp.
520 Madison Ave.
New York, NY 10022

Printed and bound by CPI Group (UK) Ltd, Croydon, CR0 4YY

03/10/2024

01040433-0007